Designing Effective Math Interventions

Design effective, learner-driven math interventions with this accessible and thought-provoking guidebook. Learn how to set up instruction to promote participation and understanding, plan purposeful and targeted tasks, develop student thinking, and create tools to assess student work in a way that measures learning, not just performance. Chapters explore questions that educators frequently struggle with when designing interventions, offering user-friendly research and evidence-based strategies to help overcome common hurdles. This book is essential reading for anyone seeking an adaptive approach to Tier 2 and 3 interventions that position struggling students as competent learners.

Jessica H. Hunt is Associate Professor of Mathematics Education and Special Education at North Carolina State University, USA. She previously taught elementary and middle school mathematics and came to love teaching students deemed to be at risk for mathematics difficulties or labeled as having disabilities.

Jenny Ainslie is a National Board Certified teacher with nearly twenty years of experience in education. She previously taught middle school mathematics and has had the privilege to hold district-level positions which have allowed her to implement and support effective mathematics instruction that is inclusive of all students at all grade levels.

Other Eye On Education Books

Available from Routledge (www.routledge.com/k-12)

Unpacking your Learning Targets: Aligning Student Learning to Standards
Sean McWherter

Coding as a Playground: Programming and Computational Thinking in the Early Childhood Classroom
Marina Umaschi Bers

Implementing Project Based Learning in Early Childhood: Overcoming Misconceptions and Reaching Success
Sara Lev, Amanda Clark, and Erin Starkey

A Focus on Addition and Subtraction: Bringing Mathematics Education Research to the Classroom
Caroline B. Ebby, Elizabeth T. Hulbert, and Rachel M. Broadhead

Implementing Systematic Interventions: A Guide for Secondary School Teams
Hank Bohanon, Lisa Caputo Love, and Kelly Morrissey

Designing Effective Math Interventions

An Educator's Guide to Learner-Driven Instruction

Jessica H. Hunt and Jenny Ainslie

Routledge
Taylor & Francis Group
NEW YORK AND LONDON

First published 2022
by Routledge
605 Third Avenue, New York, NY 10158

and by Routledge
2 Park Square, Milton Park, Abingdon, Oxon, OX14 4RN

Routledge is an imprint *of the Taylor & Francis Group, an informa business.*

Library of Congress Cataloging-in-Publication Data
A catalog record for this book has been requested.

ISBN: 978-0-367-81919-4 (hbk)
ISBN: 978-0-367-85740-0 (pbk)
ISBN: 978-1-003-01474-4 (ebk)

Typeset in Palatino
by SPi Global, India

Contents

1

How Are Mathematics Interventions Typically Defined and Conceptualized?

Historical Context

When you hear "mathematics intervention", what model of teaching comes to mind? For some, it may be the idea of re-teaching or pre-teaching the day's lesson. For others, it may be direct teaching of mathematics through the "I do, we do, you do" gradual-release model. This predictable lesson structure is built on the idea that teachers "own" expert mathematical thinking and gradually release that thinking (and the authority to employ it) to learners. Should learners not perform well in mathematics, the best option is for the teacher to explicitly model steps and thinking processes. The gradual-release model has served as the historical basis of mathematics intervention since its inception in the late 1980s, is heavily researched and validated (e.g., Chodura et al., 2015; Gersten et al., 2009; Stevens et al., 2018), and is advocated for as best practice for teaching mathematics to learners with mathematics difficulties and disabilities (MD).

Before we proceed, take the opportunity to reflect on your own perspectives about mathematics intervention by considering the information in Figure 1.1 below. Look at the intervention

Traditional Intervention Practices	Contemporary Approaches to Intervention
Learner as Receiver Learning is passive. Learners are unsure of their role in mathematics.	**Learner as Active Constructor** Learning is active. Learners believe that they are "thinkers and doers" of mathematics ("mathematicians").
Teacher as Transmitter Teacher identifies the "gaps" in learner thinking and remediates from there. Teacher is the holder of all information. The pace is set by the teacher.	**Teacher as Facilitator** Teacher understands where the learner is on possible progressions of learning and builds learner thinking from there. The pace is closely monitored by the teacher, but is determined by learner thinking.

FIGURE 1.1 Traditional and contemporary approaches to intervention

practices listed in the column on the left. Think about possible justifications for such a structure and what view of learners' mathematics and abilities the justifications may be relying upon. Then examine the intervention practices listed on the right. What similarities and differences between traditional and contemporary approaches do you notice? As you reflect upon your own perspectives of mathematics intervention, consider where they may fall on this continuum and *why*.

One justification for the widespread use of gradual-release models is cited in the research literature as a persistent "gap" in performance between learners with disabilities, which includes learners with MD, and their grade-level counterparts. For example, in the United States, the National Assessment of Educational Progress (NAEP, 2017) revealed that learners with disabilities were outperformed by their grade-level peers in mathematics: 215 vs. 244 average scale scores at fourth grade and 247 vs. 288 at eighth grade (respectively). In Figure 1.2 below, you can see that learners with disabilities at eighth grade scored similarly to their typically performing peers at fourth grade – a four-year "performance gap."

The historic focus on "gaps" in learners' knowledge lends itself well to using mathematics intervention time as a means of *remediation*. That is, intervention is used to find "gaps" in learners'

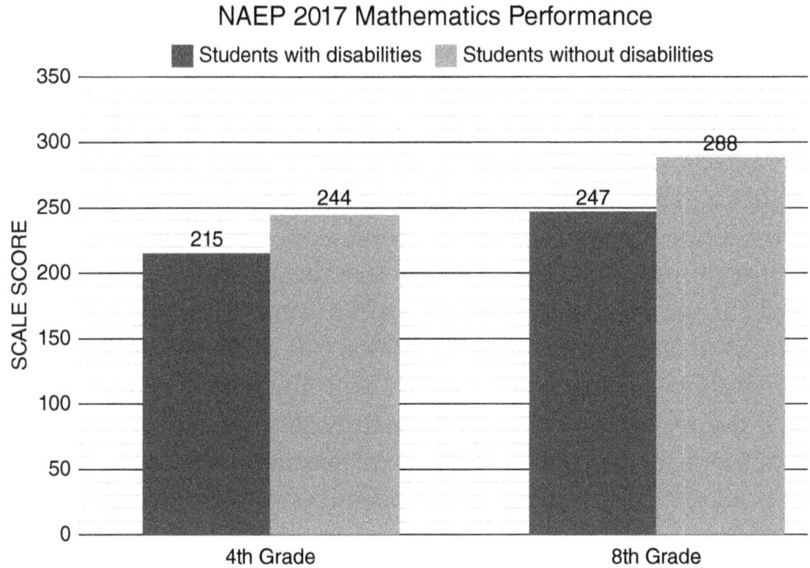

FIGURE 1.2 National Assessment of Educational Progress (NAEP) data

mathematics knowledge and fill them. The teacher's aim is to improve learners' mathematics performance, and his or her pedagogical tool is to model expert mathematical thinking and procedures for learners to acquire and practice (Fuchs et al., 2017; Geary, 2004). For learners with MD, a second aim is often to use the pedagogical tool of remediation to make up for perceived deficits that learners have in their cognition (e.g., regulatory processes and memory) – to ensure that learning is not hampered by the deficits. *Learning* in these kinds of interventions is defined in terms of *product*, or improvement in learners' responses to interventions, and is evident by increased mathematics performance after instruction has concluded.

Rethinking History

Why rethink this history of intervention, especially one that has such an evidence base behind its use? One reason, we contend, is based in the focus of products, or performance improvement, versus the processes of learning. Although increasing learners' performance in mathematics is a notable goal, defining mathematics learning solely in terms of product for learners with MD is misguided and potentially *disabling* to learners with MD. How can this be the case?

First, we argue that the depiction of cognitive difference as a deficit and as a justification for gradual release or other directive forms of pedagogy is, in short, problematic. Research has shown that seemingly deficient cognition is malleable; that is, it can grow and change over time given the opportunity to do so (Abiola & Dhindsa, 2012; Dehaene et al., 2004; Zull, 2002). Moreover, different forms of cognition coincide with the notion of *neurodiversity*, the recognition that all human brains are highly variable, that there are no typical or "normal" learners. Neurodiversity is a social justice movement that recognizes *both* the strengths and challenges of particular cognitive differences, such as learning disabilities, and thus can be used to understand MD. Learners with MD, especially learners with disabilities, often have unique or different ways of engaging with the world. We do not conceptualize these as deficits but as differences (c.f. Lewis, 2014). Mathematics intervention, then, should not aim to fix those differences but to carefully design instruction that leverages the mathematical strengths of these learners (Hunt et al., 2016, 2019, 2020; Hunt & Silva, 2020; Lewis, 2017; Lewis et al., 2020; van Garderen et al., 2020).

Second, if the modeled thinking used in remediation is that of an "expert", then only learners who can relate to the expert's thinking will make sense of it. This potentially privileges some while disenfranchising others. Research suggests that children's conceptions of mathematics are qualitatively distinct from those of adults (Steffe and Olive, 2009), often commence in non-symbolic ways that are context-dependent (Pirie & Kieren, 1994; Tzur, 2004), and can progress in ways that educators may not expect in advance of instruction (Hunt & Silva, 2020; Hunt et al., 2019; Olive & Vomvoridi, 2006; Tzur, 2004). By basing interventions solely upon adults' abstracted and "expert" frames of reference, we risk privileging limited ways of reasoning and can overlook how learners currently make sense of mathematics and strengths they can use to advance their reasoning.

In sum, we bring up the historic conversation of "performance gaps" not to perpetuate this rhetoric but to challenge and redirect it. The issue at hand is not a gap in performance but a gap in the *opportunities* that learners with MD have to know and learn mathematics through reasoning and sense-making. Overlooking the mathematical knowledge that learners with MD *do* hold places the "problem" of low performance within the learner as opposed to the lack of opportunity that, in part, may have contributed to the inequities in the first place (e.g., Balu et al., 2015).

Re-Conceptualizing Math Interventions

To re-conceptualize mathematics intervention, we need to restore opportunities that learners have to use and make sense of their own reasoning so it can advance. Part of doing that involves altering what mathematics instruction looks like during intervention time so it does not rely solely on the product of performance but also attends heavily to the *processes* of learning.

To dig into this idea a little more, let's consider a third-grade learner, "Ed". During a clinical interview, he was presented with a task that asked him to solve this problem: "Five tacos were shared equally among two people so that each person got the same amount. How much did each person receive?" (Hunt & Empson, 2015; Hunt, 2015). The problem was written in words on a piece of paper without a visual and read aloud with the learner. Linking cubes and two color counters were available if the learner wanted to use them. Here is the learner's thinking:

S This is easy. It says "the same amount" so these are the key words. [Draws five groups of two tacos]. Ten tacos.

T Ten tacos. How did you know it was ten tacos?

S I saw five and two so I drew five equal groups of two.

T You drew equal groups. OK. Tell me what the five groups represent.

S I made five equal groups of tacos.

T OK. How do your five groups of tacos relate to this problem?

S [Looks at interviewer for a bit] Well, they... I don't... My teacher didn't show me this yet.
T Hmm. What might you do if you had to share those [five] tacos [points to "five" in the written problem]? Can you picture them in front of you on a plate?
S Oh [learner grabs five linking cubes and sets them in front of him]! Like this – one for you, one for me. One for you, one for me [pauses and pushes leftover to the side].
T [Watches].
S [Looks at the interviewer for a bit] So they [one sharer] would get two and they [the other sharer] would get … [unsure] three tacos.
T What if they both wanted the same amount?
S There's not enough (Hunt & Empson, 2015, p. 214).

Many responses from the teacher are possible. One response often used in remediation is to consider the learner's performance – the "product" – in terms of what the learner *does not know*. For example, the teacher could consider that the learner did not immediately cut all of the tacos into two parts or the learner did not immediately know that each learner would receive two whole tacos and one half of the fifth taco. The teacher might also consider that the learner commented that he didn't learn that strategy yet in his classroom and, as a result, that he has no knowledge of fractions. Lastly, the teacher might consider that the learner was reluctant to share the remaining taco and was okay with handing out unequal shares, leading to a wrong answer. Together, the teacher may view the learner as lacking knowledge because his performance in the problem did not lead to a correct answer.

Now, let us further consider what this kind of response might promote the teacher to say and do next, as shown in Figure 1.3.

Attempts to remediate children's thinking in these ways is likely how many educators learned to design intervention instruction: Begin by identifying "what is wrong" and work to address the learner's incorrect answers and misconceptions through teacher modeling of the "right way." Yet we argue that responsiveness to learner thinking in these ways may work to

FIGURE 1.3 Teacher's focus

FIGURE 1.4 Learner's focus

sustain deficit views on the part of the teacher *and* the learner, which indeed can become disabling for learners. For example, consider the frame of reference of the learner who interacts (with the teacher) in this way after sharing his thinking shown in Figure 1.4.

A Different Focus

We challenge you to revolutionize how you use mathematics intervention by considering the *processes* of learning – that is, to use interventions as a mechanism to uncover a learner's mathematical strengths and build from them in the midst of instruction. Here is one powerful way to begin this work: Re-frame the question, "What does this learner not know?" or "Where are the gaps in learning?" to the far more powerful question, "What *does*

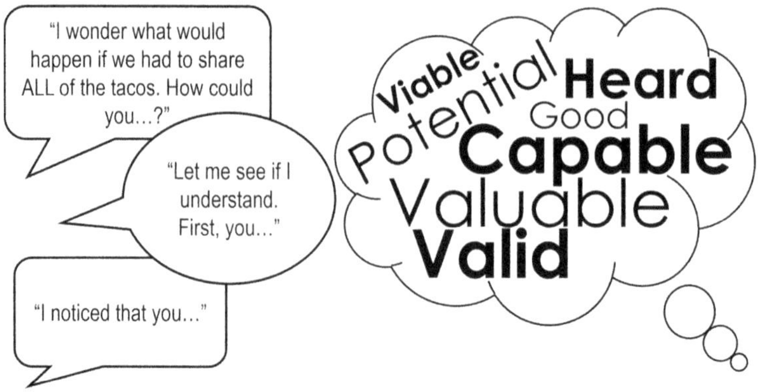

FIGURE 1.5 Revolutionized frames of reference

this learner know?" or "How is she or he thinking about it?" In doing so, you take the first step in reclaiming both the learner's potential as a mathematical thinker and doer and your own power as an educator who can support and promote learners' mathematical strengths.

From this process-oriented frame of reference, the teacher begins from the belief that the learner has a way of thinking that, from his eyes, makes sense. For example, in our example of "Ed", he knows about equality because he deals out the tacos, one by one, to the sharers, making equal groups. The teacher might also consider that, in Ed's second solution, he initially pushes the final taco to the side because he knows that he did not have enough to make equal shares in terms of whole tacos. Lastly, the teacher might consider that Ed's initial solution was to build out groups of two, so he has ideas about two as a unit that he can use to build upon in terms of fractional units. This response supports a revolutionized frame of reference for both the teacher and the learner, as shown in Figure 1.5 above.

Empowering Teachers and Their Learners

Calls for equitable mathematics learning opportunities for learners with MD to construct powerful conceptions have routinely been made over the past 30 years (National Council for Teachers of Mathematics, 2000; National Mathematics Advisory Panel, 2008; Poplin, 1988). Asking what learners *do* know and building from their strengths are promising ways forward. In this form of mathematics intervention, mathematical knowledge is not imposed on children (Baroody et al., 2004) and performance is not equated with conceptual understanding of mathematical proficiency divorced from the processes of learner learning.

Let us revisit Ed's thinking in the equal-sharing problem one more time to compare and contrast how concepts may be built up in different teacher responses. Through a lens of remediation, a teacher might respond by following through with explicit modeling of steps or the teacher might think aloud to scaffold the learner's propensity to solve the problem in a more sophisticated or desired manner. Indeed, a well-designed remediation lesson would likely result in the learner being able to reproduce the teacher's directions and increase his propensity to perform at a perceived higher level of sophistication. This approach may result in this learner successfully re-stating what was modeled for him. For example, the learner may repeat modeled thinking such as this: "Let's cut each taco into two parts because there are two people. Each one of the parts is called one half because the part repeated two times remakes a whole taco." Yet, in future problems, he may revert to what made sense to him initially (i.e., "There are not enough items to share"). Additionally, the learner may find that operating outside of the modeled strategy (e.g., using a unit fraction to create another fraction; "This picture is ½ of a whole stick of clay. Make 3/2 of the whole stick of clay") is puzzling and disconnected to how he makes sense. Or he may try to use the modeled strategy in tasks where it does not make sense to do so.

Now imagine that the teacher begins instruction with a different response. Using the learner's thinking about equality as a relation of pieces to sharers, the teacher poses a similar problem of two people sharing one taco equally (Tzur, 2007) or the same task (two people share five tacos equally) with the tacos served one at a time (Streefland, 1993). In both tasks, the learner focuses his goal on one whole and begins to consider how to use all of the whole and make equal shares because both situations work from his thinking and how he currently makes sense.

When mathematical understanding (or learning) begins with the child engaging in mathematical situations that both use and challenge his ways of knowing, it allows a space to "grapple with key mathematical ideas that are comprehensible but not yet well formed" (Hiebert & Grouws, 2007, p. 387). When teachers use a learner's strengths, he or she is positioned as already possessing a way of knowing that he or she uses to understand and to learn, building more productive dispositions and powerful conceptions in mathematics. Framing instruction from a position of "What does this learner know and how can I use it?" creates and sustains views of the child as mathematically enabled, removing the "problem" from the child and placing it, as a challenge, on the instructional design and interactions between teachers and children. *This is the aim of this book*: to empower you, the teacher, to meet this challenge.

References

Abiola, O. O., & Dhindsa, H. S. (2012). Improving classroom practices using our knowledge of how the brain works. *International Journal of Environmental and Science Education*, 7(1), 71–81.

Balu, R., Zhu, P., Doolittle, F., Schiller, E., Jenkins, J., & Gersten, R (2015). Evaluation of response to intervention practices for elementary school reading. NCEE 2016-4000. *National Center for Education Evaluation and Regional Assistance.*

Baroody, A. J., Cibulskis, M., Lai, M. L., & Li, X. (2004). Comments on the use of learning trajectories in curriculum development and research. *Mathematical Thinking and Learning*, 6(2), 227–260.

Chodura, S., Kuhn, J. T., & Holling, H. (2015). Interventions for children with mathematical difficulties: A meta-analysis. Zeitschrift fur psychologie. *Journal of Psychology*, 223(2), 129–144. https://doi.org/10.1027/2151-2604/a000211

Dehaene, S., Molko, N., Cohen, L., & Wilson, A. J. (2004). Arithmetic and the brain. *Current Opinion in Neurobiology*, 14(2), 218–224.

Fuchs, L. S., Malone, A. S., Schumacher, R. F., Namkung, J., & Wang, A. (2017). Fraction intervention for students with mathematics difficulties: Lessons learned from five randomized controlled trials.

Journal of Learning Disabilities, 50(6), 631–639. https://doi.org/10.1177/0022219416677249.

Geary, D. C. (2004). Mathematics and learning disabilities. *Journal of Learning Disabilities*, 37(1), 4–15.

Gersten, R., Beckmann, S., Clarke, B., Foegen, A., Marsh, L., Star, J. R., & Witzel, B. (2009). *Assisting students struggling with mathematics: Response to Intervention (RtI) for elementary and middle schools* (NCEE 2009-4060). Washington, DC: National Center for Education Evaluation and Regional Assistance, Institute of Education Sciences, U.S. Department of Education. Retrieved from http://ies.ed.gov/ncee/wwc/publications/practiceguides/.

Hiebert, J., & Grouws, D. A. (2007). The effects of classroom mathematics teaching on students' learning. In F. K. Lester (Ed.), *Second handbook of research on mathematics teaching and learning*) (pp. 371–404.Charlotte, NC: Information Age Publishing.

Hunt, J. H., & Empson, S. B. (2015). Exploratory study of informal strategies for equal sharing problems of students with learning disabilities. *Learning Disability Quarterly*, 38(4), 208–220.

Hunt, J. (2015). How to better understand the diverse mathematical thinking of learners. *Australian Primary Mathematics Classroom*, 20(2), 15–21.

Hunt, J. H., Welch-Ptak, J. J., & Silva, J. M. (2016). Initial understandings of fraction concepts evidenced by students with mathematics learning disabilities and difficulties: A framework. *Learning Disability Quarterly*, 39(4), 213–225.

Hunt, J. H., Silva, J., & Lambert, R. (2019). Empowering students with specific learning disabilities: Jim's concept of unit fraction. *The Journal of Mathematical Behavior*, 56, 100738.

Hunt, J. H., Martin, K., Khounmeuang, A., Silva, J., Patterson, B., & Welch-Ptak, J. (2020). Design, development, and initial testing of asset-based intervention grounded in trajectories of student fraction learning. *Learning Disability Quarterly*, 0731948720963589.

Hunt, J., & Silva, J. (2020). Emma's negotiation of number: Implicit intensive intervention. *Journal for Research in Mathematics Education*, 51(3), 334–360.

Lewis, K. E. (2014). Difference not deficit: Reconceptualizing mathematical learning disabilities. *Journal for Research in Mathematics Education*, 45(3), 351–396.

Lewis, K. E. (2017). Designing a bridging discourse: Re-mediation of a mathematical learning disability. *Journal of the Learning Sciences*, 26(2), 320–365.

Lewis, K. E., Sweeney, G., Thompson, G. M., & Adler, R. M. (2020). Integer number sense and notation: A case study of a student with a mathematics learning disability. *The Journal of Mathematical Behavior*, 59, 100,797.

National Assessment of Educational Progress (NAEP). (2017). *National Center for Education Statistics*, Retrieved from https://www.nationsreportcard.gov/math_2019/nation/achievement?grade=4. Retrieved September 30, 2019.

National Council for Teachers of Mathematics (2000). National Council of Teachers of Mathematics. *Principles and standards for school mathematics*. Reston, VA: National Council of Teachers of Mathematics.

National Mathematics Advisory Panel. (2008). *Foundations for success: The final report of the National Mathematics Advisory Panel*: US Department of Education.

Olive, J., & Vomvoridi, E. (2006). Making sense of instruction on fractions when a student lacks necessary fractional schemes: The case of Tim. *The Journal of Mathematical Behavior*, 25(1), 18–45.

Pirie, S., & Kieren, T. (1994). Growth in mathematical understanding: How can we characterise it and how can we represent it?. In P. Cobb (Ed) *Learning mathematics* (pp. 61–86). Dordrecht: Springer.

Poplin, M. S. (1988). Holistic/constructivist principles of the teaching/learning process: Implications for the field of learning disabilities. *Journal of learning Disabilities*, 21(7), 401–416.

Steffe, L. P., & Olive, J. (2009). *Children's fractional knowledge*. New York, NY: Springer Science & Business Media.

Stevens, E. A., Rodgers, M. A., & Powell, S. R. (2018). Mathematics interventions for upper elementary and secondary students: A meta-analysis of research. *Remedial and Special Education*, 39(6), 327–340. https://doi.org/10.1177/0741932517731887

Streefland, L. (1993). Fractions: A realistic approach. In T.P. Carpenter, E. Fennema, & T. Romberg (Eds.) *Rational numbers: An integration of research* (p. 289–325). New York, NY: Routledge.

Tzur, R. (2004). Teacher and students' joint production of a reversible fraction conception. *The Journal of Mathematical Behavior*, 23(1), 93–114.

Tzur, R. (2007). Fine grain assessment of students' mathematical understanding: Participatory and anticipatory stages in learning a new mathematical conception. *Educational Studies in Mathematics*, 66(3), 273–291.

van Garderen, D., Lannin, J. K., & Kamuru, J. (2020). Intertwining special education and mathematics education perspectives to design an intervention to improve student understanding of symbolic numerical magnitude. *The Journal of Mathematical Behavior*, 59(100): 782.

Zull, J. E. (2002). *The art of changing the brain: Enriching teaching by exploring the biology of learning*. New York, NY: Stylus Publishing, LLC.

2

What Does It Mean For a Learner to Know and Learn Mathematics Well?

Shifting from Products to the Processes of Learning

In Chapter 1, we focused on the limitations of remediation as a productive means of intervention. We linked remediation to opportunity gaps for learners to learn from their own ways of making sense. Our aim in this book is to support teachers to understand how learners think so that instruction can be based on that thinking. To reduce the opportunity gap, we emphasize a need to use core instructional mechanisms that allow teachers first to seek to understand the knowledge that learners already have and then to work from that knowledge to design instruction.

To do this work, teachers need to attend to the *processes* of learning – that is, attend to learners' mathematical thinking – as a basis for instructional change. In a nutshell, if learning is defined by the process of learners' growing mathematical thinking as opposed to the product of performance, then we can design instruction to not only work from the learners' thinking but support its advancement over time. This chapter will set the stage to help us focus on the learning process. Let's begin by defining what it means for a learner to *know* mathematics well.

Knowing Mathematics

In 2009, Paul Lockhart wrote:

> By concentrating on *what*, and leaving out *why*, mathematics is reduced to an empty shell … If you deny learners the opportunity to … pose their own problems, make their own conjectures and discoveries, to be wrong, to be creatively frustrated, to have an inspiration, and to cobble together their own explanations and proofs – you deny them mathematics itself.
>
> (p. 5)

His statement is compelling because it connects to how we will describe what it means to *learn* mathematics well in later chapters. For now, we will connect Lockhart's words to an illustration of one learner with learning disabilities, "Emma", a friendly 10-year-old learner who often generated creative ways to solve problems and who came to *know* mathematics well (see Hunt & Silva, 2020).

In order to do this, we will be referencing Kilpatrick et al. (2001), who describe *mathematical proficiency* as encompassing five distinct strands: *conceptual understanding*, *procedural fluency*, *strategic competence*, *adaptive reasoning*, and *productive disposition* (as shown in Figure 2.1). Although all strands are necessary for proficiency, we tell Emma's story through the strands of conceptual understanding, productive disposition, and procedural fluency.

Conceptual understanding is defined by the authors as a "comprehension of mathematics concepts, operations, and relations" (Kilpatrick et al., 2001, p. 5). When we started working with Emma, her conceptual understanding was initial. She often resorted to procedures that she applied incorrectly or in the wrong context. For example, in problems that involved a missing addend (e.g., "I have 14 total cubes. Six are showing and the rest are hidden. How many are hidden?"), Emma would add the total of 14 and the addend of six together in a column, state

The Five Strands of Mathematical Proficiency	
Conceptual Understanding	comprehension of mathematical concepts, operations, and relations
Procedural Fluency	skill in carrying out procedures flexibly, accurately, efficiently, and appropriately
Strategic Competence	ability to formulate, represent, and solve mathematical problems
Adaptive Reasoning	capacity for logical thought, reflection, explanation, and justification
Productive Disposition	habitual inclination to see mathematics as sensible, useful, and worthwhile, coupled with a belief in diligence and one's own efficacy.

FIGURE 2.1 Five interconnected strands of mathematics proficiency (Kilpatrick et al., 2001)

the answer as 20, and frown. In another problem, of combining 18 cubes and six cubes, Emma would count out 18 cubes, then count out six cubes, and then count all of them starting from one to determine the total (called a "count-all strategy"). When asked about her thinking, Emma stated that the "1" in 18 meant one as a single unit as opposed to a unit of ten. In fact, Emma solved all addition problems with a count-all strategy or a procedure not yet related to a concept of ten as a unit. Here is her thinking when asked to add 3 and 13:

S Like 3 …4, 5, 6…16. And then the one.
T You said, 3…4, 5, 6 and the one. What's the one?
S The one is the… [pauses, frowns].
T Can you show me what you mean? [hands child a paper and pen]
S So it was…13 plus 3 is… [writes 13 + 3 = 16 long form].
T [points to the one in 13] So this is the one?
S That's the one in the 13.
T But is that like, ONE [shows one finger]?
S Yeah only one.
T So, if this is one, how come I can't just count one more…seven?
S Because…three doesn't have a one.

T Suppose you had started counting from this three [points to the three underneath 13 and references third space on the game board]? How would you do that? [removes paper and game board].

S [puts up hand] 3…, 4 [raises one finger], 5 [raises 2nd finger], 6 [3rd finger], 7 [4th finger], … 8 [5th finger], 9 [6th finger], 10 [7th finger], 11 [8th finger], 12 [9th finger], 13 [10th finger; stares at ten fingers and pauses for 5 seconds].

T 13… How many have you counted so far?

S Ten.

T How many more do you need to count?

S I think… [sticks out lower lip, pauses for 3 seconds]. I think… [frowns, looks down].

T [grabs a paper and pen] So you started at three [writes three and an empty number line] …and you did 4 [makes hop on number line], 5 [makes hop on number line], 6 [makes hop on number line], 7 [makes hop on number line], 8 [makes hop on number line], 9 [makes hop on number line], 10 [makes hop on number line], 11 [makes hop on number line], 12 [makes hop on number line], 13 [makes hop on number line]. Then you stopped and did this [holds up all ten fingers and wiggles them, covers up the number line].

S Oh! 13… 14 [raises a finger], 15 [raises a finger], 16 [raises a finger]. I need three more. I needed three more to get 16.

T How did you know it was three more?

S Because my answer got me 16.

T OK. I wonder if there is a way to figure it out without having to know the answer first.

S [shrugs].

Yet, as we worked with Emma, her concept of number became quite sophisticated! In fact, after less than three hours of intervention based on developing numbers as composite units (i.e., a view of numbers, such as 10, as a composition of other numbers, such as 8 and 2 or 6 and 4), Emma successfully began to build number as a composite unit. Over time, Emma also used a more efficient set of strategies to operate with. For example, if asked a problem about a missing addend (e.g., "I have 11 cubes in total.

Four are showing and the rest are hidden. How many are hidden?"), Emma would write 4 + ____ = 11, count up from 4 using one hand (e.g., 5-6-7-8-9, 10-11), say that the answer is seven, and show seven using five and two on her hands. Other times, she connected the use of an algorithm to consider 18 – __ = 10 through considering the parts of eighteen (i.e., 8 and 10), no longer claiming the one in 18 is "one" but that it represents "one ten." Finally, when asked a question about composing the number 12 in various ways, Emma wrote 8 + 4 then 4 + 8 and said they were equal. She then generated additional solutions (e.g., 3 + 9) and explained with words and number sentences how they were related to 12.

The changes evident in Emma's reasoning also connect to procedural fluency. *Procedural fluency* is defined as "skill in carrying out procedures flexibly, accurately, efficiently, and appropriately" (Kilpatrick et al., 2001, p. 5). When we began working with Emma, procedures were something that she drew upon yet her use of and facility with procedures were limited. For example, Emma used procedures yet did so with incomplete or inaccurate beliefs (e.g., the "1" in 18 represents a single unit). In this way, we contend that procedural fluency can be further understood as routine expertise and adaptive expertise (Baroody, 2011).

Routine expertise is when learners master procedures to become highly efficient and accurate. *Adaptive expertise* intertwines with conceptual understanding in that it allows the learner to connect new solutions to problems and even new procedures for solving problems. Emma progressed from using an incomplete version of procedural routine expertise to the beginnings of adaptive expertise. For example, Emma created a new strategy for adding, constructing a count-on and a break-apart strategy to consider addition and subtraction. The progression highlights the interconnected nature of her newly built conceptions with the advances in procedural fluency.

Productive disposition is defined as "habitual inclination to see mathematics as sensible, useful, and worthwhile, coupled with a belief in diligence and one's own efficacy" (Kilpatrick et al., 2001, p. 5). The beliefs a learner might have for themselves in their own mathematical potential have lasting implications both in terms of mathematics proficiency and in their propensity to choose math-related careers (Bittinger, 2018). At the start of our work with Emma, we did not measure her productive dispositions about mathematics directly. Yet her timid nature and pensiveness regarding her own reasoning were telling. Quite quickly, however, she went from being uncertain about her own thinking to working excitedly on the tasks, taking pride and being confident in her own answers, eagerly engaging in solving challenging tasks on her own, and explaining her work to the teacher.

From Knowing to Learning Mathematics

The strands of mathematics proficiency illustrate what it means to *know* mathematics well and are deeply connected to what it means to *learn* mathematics well. Learners become proficient in mathematics by (a) using the ideas they already have, (b) actively thinking through a problem, and (c) reflecting on their thinking and that of others. From our view, it is through this process that learners begin to build conceptual understanding, connect procedural fluency and adaptive expertise, develop strategic competence, and see themselves as "thinkers and doers" of mathematics. We cannot stress enough that *learners* need to do cognitive work during intervention time: mathematics proficiency is best supported and advanced through *learners' own ways of reasoning and sense-making*. The role of the teacher is to understand learners' prior knowledge and how mathematical thinking advances so that the teacher can support learners to advance their own reasoning. In other words, learners must build mathematics proficiency *for* themselves yet absolutely not *by* themselves.

The next part of this chapter unpacks learning as a process of three core mechanisms: (a) being active in a problem, (b) using prior knowledge, and (c) reflecting on action over time. This is *not* the same thing as learners "discovering" the mathematics on their own; indeed, the role of the teacher in this process is critical. Above, we shared Emma's initial understanding as well as how she grew to know mathematics well. What we did not illustrate is how we used the knowledge of these core mechanisms to promote her growth. We discuss these below; Chapter 3 connects specific teaching moves to these three core mechanisms of learning.

Learning Mechanism: Active Problem-Solving

To learn, learners must engage with and be *active* within mathematics (von Glasersfeld, 1995). In school settings, the learning environment (classroom or small group) is "where" learners engage and tasks become "what" the learners engage with. The learning environment and tasks must place learners' thinking and ideas at the forefront. Yet, too often, they do not: as we shared in Chapter 1, learners often become passive because the cognitive work is done by the teacher or the environment is not set up with the expectation that learner thinking is valued and will be promoted.

Removing the cognitive work from learners often perpetuates or leads to *misconceptions* (we prefer the term partial conceptions) that have implications for mathematics proficiency in learners' later lives (e.g., Hunt et al., 2019). The case of Emma is a prime illustration of a learner who had been taught in intervention environments where the cognitive work was done by the teacher and learners were not expected or invited to use their own reasoning within a problem. For example, consider Emma's thinking from the excerpt we listed earlier as she worked with us in an early tutoring session to add 3 + 13, which shows another instance of her thinking of a digit "1" in the tens place as a single unit of one:

S 16.
T Ok. How'd you get 16?
S Like 3…4, 5, 6... And then the one.
T You said, 3…4, 5, 6 and the one. What's the one?
S The one is the… [pauses, frowns].
T Can you show me what you mean? [hands child a paper and pen].
S So it was…13 plus 3 is… [writes 13 + 3 = 16 vertically].
T [points to the one in 13] So this is the one?

S That's the one in the 13.
T But is that like, ONE [shows one finger]?
S Yeah, only one.
T So, if this is one, how come I can't just count one more…seven?
S Because…three doesn't have a one.

(Hunt and Silva, 2020)

Emma had viable strategies to use within problems such as these. In fact, she began to use counting on – a strategy that she did not yet use when we began working with her – to support her procedure of adding three to the three in 13. Yet Emma then employed a procedure that was taught to her that, while affording her the correct answer, shut down her cognitive action in the problem and illustrated a partial conception that she had about the value of the digit in the tens place in the number 13.

A learner's mathematics can return to its pre-intervention state when his or her ideas are not used as a basis for instruction. For Emma, this was illustrated by her use of a procedure to get an answer that did not make sense in terms of place value or the composite structure of 13 being equivalent to 10 and three. It is also important to note that Emma's conceptual frame of reference around mathematical understanding remained largely dormant despite her receiving ongoing mathematics interventions that addressed place-value concepts. To change the outcome for Emma and other learners, how might we create an environment where they become active problem solvers?

Learning Mechanism: Use Prior Knowledge in a Problematic Situation

Learning environment and tasks should be designed to bring forward what learners already know – their *prior knowledge* – and promote them to use it (or expand it) in a problematic task. Let's first unpack what we mean by prior knowledge. *Prior knowledge* is a term we use both broadly and specifically. Because knowledge is contextual as well as cognitive, the learning environment and tasks should be designed to support learners to bring forward their entire experience to consider problems. When we work with learners, we gather information about their specific prior knowledge related to mathematics as well as broad prior knowledge of the learner as a human being. For example, when we began working with Emma, we knew she liked to play board games at home, enjoyed playing outside (e.g., hopscotch), and spent time with her mother at her job in a neighborhood bakery (broad contextual prior knowledge related to Emma as a human being). We also knew that Emma mainly used a count-all strategy along with some procedures to reason about addition (specific prior knowledge related to mathematics). Uncovering prior knowledge involves considering *all* of the learner's knowledge when designing intervention to support advancements in reasoning.

Combining what is known about learners' broad and specific prior knowledge to design interventions creates a great way for learners to be *active in problematic situations*. For example, one task we designed for Emma involved rolling a die and moving a marker across a linear number game board (Tzur & Lambert, 2011). The object of the game was for Emma to tell "How Far From the Start" she was after combining two addends: one rolled by the teacher and another by Emma. She could count all of the spaces by ones or other numbers. She could also use one of the addends to count on from. Emma played several rounds of the game in which she worked with multiple addends to arrive at a total. Figure 2.2 shows an example game board.

In the game, we found that Emma had another strategy for adding: finger counting. The teacher had just rolled a seven, and Emma rolled a nine. Figure 2.3 shows the board game play;

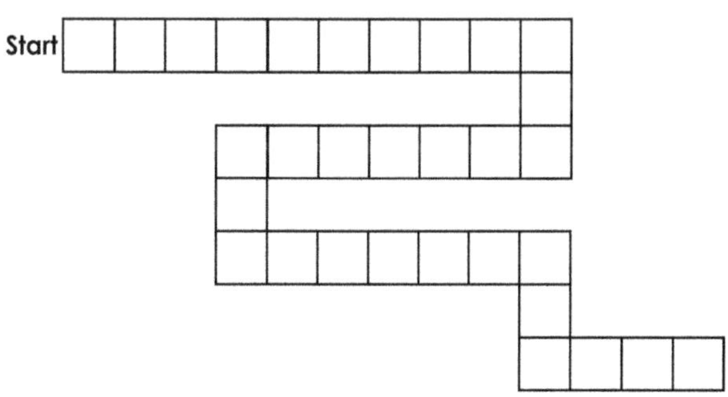

FIGURE 2.2 Example game board

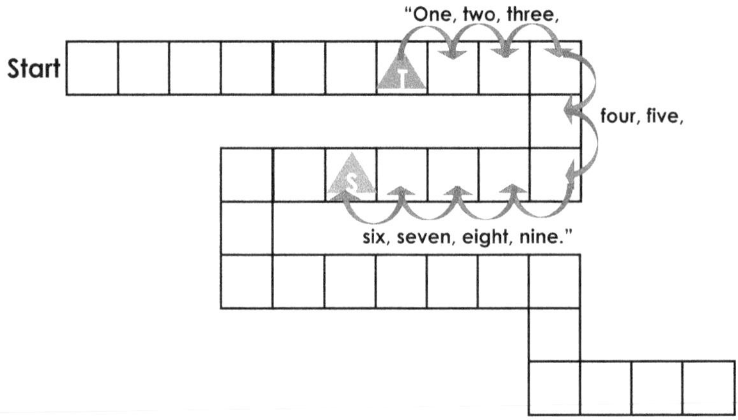

FIGURE 2.3 Emma's counting. S, student; T, teacher

below it, we summarize the conversation between the learner and the teacher.

> The teacher rolled a seven and asked for Emma's help to count out the spaces. To move the marker and count the spaces, the learner counted by twos and ones. Next, Emma rolled a nine. She counted by twos five times. With a prompt from the teacher ("How many were we trying to go?"), she recounted by twos and ones. Then, the teacher asked Emma how far from the start she was. To figure it out, the learner counted the visible game board spaces, starting from the first space, "1, 2, 3, 4, 5, 6, 7, 8, 9, 10, 11, 12, 13, 14, 15, 16."

There was intentionality when selecting this activity for Emma. First, it was a board game – something Emma enjoyed and was familiar with – that allowed her access to both her broad human and mathematical prior knowledge. She could use her count-all strategy to count out each addend and then recount the visible spaces as a strategy to solve the problems. If she had wished, we could have played the game outside with chalk to create a large (life-size) version of the board game to connect to Emma's love of hopscotch. Second, the activity allowed us to create a *problematic situation* for Emma. Generally, this means we could introduce constraints (e.g., covering of addends/board game spaces and using larger numbers) that might promote Emma to change her strategy.

The activity brought forward Emma's prior knowledge. To make it problematic, we used a constraint, as shown in Figure 2.4. That is, we began to cover the visible spaces after Emma counted them. When we did, Emma brought forward a new strategy as she added the addends of nine and six. We summarize the conversation between the learner and the teacher below.

Emma counted out each person's rolls, or the problem "addends", by counting the visible game spaces by threes. Next, the learner wiggled her fingers under the table after the teacher asked her how far from the start she was on the game board. The teacher encouraged Emma to explain or show her thinking; at that point, Emma showed how to count on, one at a time, from nine, raising one finger every time that she counted by one. It was interesting that the learner counted by ones on one hand; when she ran out of fingers, she closed her hand and began to count by ones again using the same hand. When the teacher asked Emma how she knew when to stop counting, Emma said, "I don't know." At that point, the teacher "re-showed" the learner's finger counting and repeated her question to Emma ("How did you know when to stop counting?"). Emma then explained that she knew when to stop counting because five plus one was equal to six.

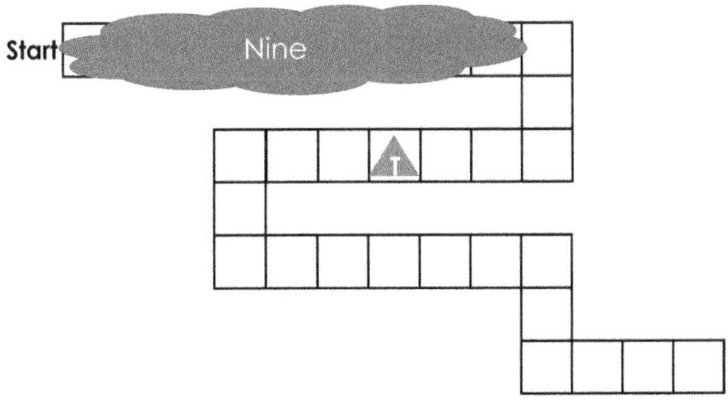

FIGURE 2.4 Covering addends. T, teacher

Covering the nine made the addition problematic for Emma. That is, she could no longer count the spaces starting at zero to figure the sum. As a result, she created a new strategy, a *count on*, that supported the use of her fingers to stand in for the board game spaces she could no longer see. When confronted with the need to do so, learners can advance their reasoning.

Learning Mechanism: Noticing and Reflecting on Actions and Results

Reflection on problematic action in a task is fundamental to learning and can occur in two forms. One form of reflection occurs when learners *notice* what happens as a result of their actions as they work in and across relevant situations. In Emma's case, she may have noticed that "Counting all of the objects helped me find 'how many'." Generally, learners can become aware of and adapt their ways of knowing through this kind of reflection with or without support from teachers. For instance, with our problem constraint, Emma may have noticed, "If I use the first addend as a starting point, I can also reach my goal and I don't have to count all."

The second form of reflection occurs through *repeated reflection* on connected and distal experiences. In order to expand their reasoning, learners need to reflect upon the consistent results of their actions. After playing this game several times, Emma may have thought, "I need to keep track of the second addend", and built up a way to do so through her finger counting. As learners engage and reflect across repeated experiences, they begin to anticipate results before they solve problems, which helps their knowledge become more efficient and abstract. For example, as we continued to work with Emma, we imposed more problem constraints, such as using larger numbers, covering both addends, and stepping away from the context of the game board entirely. Over time, Emma began to break larger addends apart (13 as ten and 3), explained the necessity of the anticipated results of her actions (e.g., "10 and 3 are inside of 13"), and realized why that result is important, useful, and mathematically justified (e.g., "Wholes can be composed and decomposed into parts").

Look again at the explanation of how Emma added nine and six within the game context we described above. This time, we note in **bold** those instances of Emma's noticing and reflection and in *italics* those questioning strategies that the teacher used to support it:

> Emma counted out each person's rolls, or the problem "addends", by counting the visible game spaces by threes. Next, the learner **wiggled her fingers under the table** after the teacher asked her how far from the start she was on the game board. The teacher encouraged Emma to *explain or show her thinking*; at that point, **Emma showed how to count on, one at a time, from nine, raising one finger every time that she counted by one. It was interesting that the learner counted by ones on one hand; when she ran out of fingers, she closed her hand and began to count by ones again using the same hand**. When the teacher asked Emma *how she knew when to stop counting*, Emma said, "I don't know." At that point, *the teacher "re-showed" the learner's finger counting and repeated her question to Emma ("How did you know when to stop counting?")*. Emma then explained that she knew when to stop counting because five plus one was equal to six.

If teachers view "intervention" as a platform for learners to gain access to their own mathematics and reflect upon it as mathematics learners, then they will be empowered. A core aspect of this work involves designing intervention based upon *learning*. Learning happens by *(a) being active problem solvers, (b) using their prior knowledge in problematic situations, and (c) noticing and reflecting on results of their mathematical actions*. The next chapters connect teaching moves to support these three mechanisms of learning within intervention time.

References

Baroody, A. J. (2011). Learning: A framework. In F. Fennell (Ed.), *Achieving Fluency: Special Education and Mathematics* (pp. 15–58). Reston, VA: National Council of Teachers of Mathematics.

Bittinger, J. (2018). STEM pipeline for students with disabilities: From high school to intentions to major in STEM. Doctoral Dissertations. 1313.

Hunt, J. H., MacDonald, B. L., & Silva, J. (2019). Gina's mathematics: Thinking, tricks, or "teaching"?. *The Journal of Mathematical Behavior*, 56, 100707.

Hunt, J., & Silva, J. (2020). Emma's counting and coordination of units: Supporting sense making for students with working memory differences. In Sam Otten, A. Candela, Z. de Araujo, C. Haines, & C. Munter (Eds.) *Proceedings of the Forty-first Annual Meeting of the North American Chapter of the International Group for the Psychology of Mathematics Education* (pp. 1477–1481). St. Louis, MO: University of Missouri.

Kilpatrick, J., Swafford, J., & Findell, B. (Eds.). (2001). *Adding it Up: Helping Children Learn Mathematics*. Washington, DC: National Academy.

Lockhart, P. (2009). *A Mathematician's Lament: How School Cheats us out of our most Fascinating and Imaginative Art Form*. New York, NY: Bellevue Literary Press.

Tzur, R., & Lambert, M. A. (2011). Intermediate participatory stages as Zone of Proximal Development correlate in constructing counting-on: A plausible conceptual source for children's transitory 'regress' to counting-all. *Journal for Research in Mathematics Education,* 42(5), 418–450.

von Glasersfeld, E. (1995). *Radical Constructivism: A Way of Knowing and Learning*. Washington, DC: Falmer.

3

How Can Teaching Be Designed to Bring Forward Learners' Prior Knowledge?

Connecting Teaching to Learning: The Importance of Prior Knowledge

The goals of any instructional time with a learner, especially in an intervention setting, are to build relationships and promote learning. You probably know from your own experiences that neither is an easy process or a linear one. We might prefer to have a list of ten steps that we follow, and check them off as we go, as opposed to the more intricate work of getting to know learners' multiple sources of mathematical thinking and overall ways in which they build meaning. Yet, as we explained in Chapters 1 and 2, the work of teaching, especially teaching with the goal of promoting learning and relationships, is not that simple. To begin this work, educators can link teaching to how learning happens (see Chapter 2) so that each learner's prior knowledge, actions in a task, and reflection upon the mathematics they are working on can be brought forward and promoted while building rich mathematics proficiency.

The focus of this Chapter is on the intentional role of the teacher within intervention spaces with the goal of learning. One thing to keep in mind is that we situate this work as if you do not yet know your learners or the mathematics they are bringing into the classroom. This is done to support teachers who are:

- Just starting the school year in a variety of learning spaces
- Pushing into a whole class setting and supporting mathematics learning in small group time
- Meeting with small groups of learners just outside the whole classroom space in small environments to further promote learner learning.

"Small Environments"

We cannot begin the chapter without unpacking what we meant in the preceding paragraph by "Small Environments." One way to define Small Environments is by the size of the instructional group. Intervention groups are often small (i.e., 3 or 4 learners), as the smaller ratio of learners is found to be beneficial for learner ← → teacher and learner ← → learning and interaction (e.g., Fuchs et al., 2014). Another way to define Small Environments is by how we position learners within intervention spaces and consider their potential for learning. For us, Small Environments are places where we, as teachers, know that all learners possess a way of knowing mathematics. If learners bring forward knowledge that we do not expect, that knowledge, like the learner, is valid (as opposed to deficient). Learners can be thought of as little mathematicians who learn by connecting what they already know to results of their actions through reflection and conversation. Our job is to listen to learners' reasoning, figure out how they understand, and build on that understanding.

Overview of Key Teaching Moves

In Chapter 2, we defined learning in three ways: (a) using the ideas that a person already has, (b) actively thinking through a problem, and (c) reflecting on their own thinking and making connections with the thinking of other people. Below, we will

TABLE 3.1 Connecting teaching to learning

Learning mechanisms (learner actions)	*Key teaching move (teacher actions)*
Using ideas that one already has	#1: Bringing forward learners' prior knowledge
Actively thinking through a problem	#2: Promoting *learners'* participation and thinking and the need for change
Reflecting on thinking and how it connects to other's thinking	#3: Promoting noticing, reflecting, and connections through purposive teaching

provide an overview of how teaching can connect to each of these three mechanisms, as shown in Table 3.1. This chapter will focus on practical suggestions and tools you may use in order to investigate and bring forward learners' prior knowledge. In the chapters that follow, the other two key teaching moves will be unpacked with suggestions to support the work of teaching in Small Environments.

Key Teaching Move #1: Bring Forward Learners' Prior Knowledge

Combining what is known about learners' broad and specific prior knowledge is fundamental to designing interventions. In Chapter 2, we defined prior knowledge both broadly and more specifically as it relates to mathematics. It is not uncommon in the intervention world to rely on computer-based diagnostic assessments in order to determine a learner's prior knowledge or "where to begin" with mathematics instruction or both. However, we contend that a computer is only one tool that teachers can use when it comes to uncovering the conceptions of learning that a learner may hold. Many others are necessary to foster learning as well as build the relationships that help learners to see themselves as capable in mathematics (McLeskey et al., 2017). Three interconnected tools can help you to uncover not only learners' specific prior knowledge related to mathematics but also the broader informal knowledge bases that learners can and will draw upon to make sense.

The first tool is observation. Observation is used so that teachers become "learners of their learners." That is, part of the basis of Small Environments is knowing your learners – what is meaningful to them, what spaces they feel competent within, and how that could relate to mathematics. Observations can take place in many settings: lunch, other classes, recess, and the humanities. Learners should always understand the purpose for the observations and agree to their being conducted.

The second tool is conversation. Conversations can be informally based or linked to discovering learners' competencies, their interests, their homes and neighborhoods, and their dispositions related to mathematics. For us, having conversations with learners – either about who they are or about mathematics – is a powerful way to uncover their prior knowledge.

The third tool is tasks. Tasks can act as platforms in which to discover and bring forward learners' broad and specific prior knowledge. They can be designed widely to ensure access and determine learners' prior mathematical thinking, or they can be designed to bridge prior knowledge to new and connected concepts. In the last sections of this chapter, we will discuss these interconnected tools and introduce practical ways to use them in school settings.

Using Observation, Conversation, and Task-Based Conversation to Bring Forward Prior Knowledge

To begin to consider how observation, conversation, and the design of tasks work to promote or shut down access to a learner's prior mathematical knowledge, step into the shoes of Jamar, a sixth-grade learner who was learning about ratio and proportion. Jamar takes part in mathematics intervention time every day.

> You have come to dread this time of the day, not because you do not enjoy mathematics but because you do not enjoy **this** mathematics. Every day, you are instructed to solve problems in ways demonstrated to you during the first part of the lesson and are corrected by the teacher when your thinking deviates from the demonstrated

> solution strategy. Today, you are working on ratio and proportion. You are working to correctly solve problems such as $\frac{3}{4} = \frac{x}{28}$ and have been shown a method to solve it. The method involves multiplying the numerator of the fraction on the left-hand side of the equals sign with the denominator on the right-hand side of the equals sign and dividing by the denominator on the left-hand side of the equals sign ("Cross multiply and divide," the teacher keeps saying). Although you are able to replicate the demonstrated method, you find yourself becoming bored, so you begin to draw out groups of three and four until you have seven groups of four and seven groups of three. You knew previously that four and 28 are related and you make a connection between four, 28, and the seven groups of four in your drawing. Yet, before you can think about the connections to your representation further, you are asked to abandon that thinking and move toward the demonstrated strategy. When the next problem ($\frac{2}{3} = \frac{x}{4.5}$) is shown, you put your head down on the desk and begin to count the minutes until you can go to one of your favorite parts of the school day: lunch. To drink, you have a water bottle and a packet of iced tea mix that your dad brought you from his restaurant and you've been considering how much of the mix to use to make the tea taste right.

Jamar had viable strategies for considering the problems that were shut down by the intervention because it did not consider Jamar's prior knowledge – both his specific mathematical strategies and his broader knowledge brought by his interests and home environment. Designing instructional experiences that support learners to access mathematics from their own ways of knowing *from the start* (CAST, 2018) can promote learners' disposition and development of conceptual understanding, reasoning, sense-making, and adaptive expertise (National Research Council and Mathematics Learning Study Committee, 2001). Yet, too often, interventions are grounded in tasks or means of learner engagement that do little to honor the many ways that learners may already think about the concept addressed in the intervention. That is the point of the remainder of this chapter: to support you to uncover your learners' prior knowledge so that you can bring it forward and use it in the Small Environment.

Below, we will expound upon observation, conversation, and tasks, supplying tools that you can use for each area. At times, we will use a real-world example to help illustrate the use of the tools introduced as well as the specifics of the dance between teacher moves and learner action in bringing forward prior knowledge. Specifically, we will tell the story of "Stu", a fifth-grade learner with a learning disability whom we worked with in both multiplicative reasoning and fractional reasoning (Hunt et al., 2016).

Observation

As we mentioned above, observation is used so that teachers become "learners of their learners." You can use observation to get to know your learner a bit better. Ask that learner whether you can "shadow" them for part of a day. Tell them that you would like them to show you what they are good at doing, what they enjoy, and other things important to them on a regular school day. Think about convenient times for the observation (e.g., at lunch, recess, and in other classes) and be sure to ask the learner whether it is okay to shadow them during these times. Below are some considerations for your observation from the work of TEACH MATH (Drake et al., 2015; see https://teachmath.info/):

- If you are observing the learner's classes, when does the learner participate in the class, and when does the learner refrain from participation? Is the learner's participation different in different classes?
- When you are listening to the learner talk about themselves or their peers talking to or about the learner, does it seem like the learner is "known for" anything in particular (e.g., good athlete, good at math, good reader, artist, "popular" with other kids, in a certain "group")? (Here, you are observing for the learner's perceived "identity.")
- What does school look like from the learner's perspective?

- In what ways is the learner showing competence (i.e., skill, ability, knowledge, creativity, problem-solving, and reasoning) in various contexts? Pay particular attention to ways you see the learner showing his or her competence that might not be captured by a teacher or by a typical curriculum task.

Informal Interview

There are several aspects of a learner's broader informal knowledge that can be uncovered through a short informal interview. An informal interview is a way to get to know more about your learner's home and community lives so that you can use them within small intervention environments (Drake et al., 2015). During an informal interview, a teacher might find out more about the learner's interests, activities that the learner engages in within their community, and what the learner identifies as activities in which they excel. Teachers may also identify locations in the learner's neighborhood or surrounding the school that the learner finds familiar. These locations may have potential contexts for mathematical connections. Finally, informal interviews help teachers find out more about the learner's ideas, attitudes, and dispositions toward mathematics. Figure 3.1 lists questions

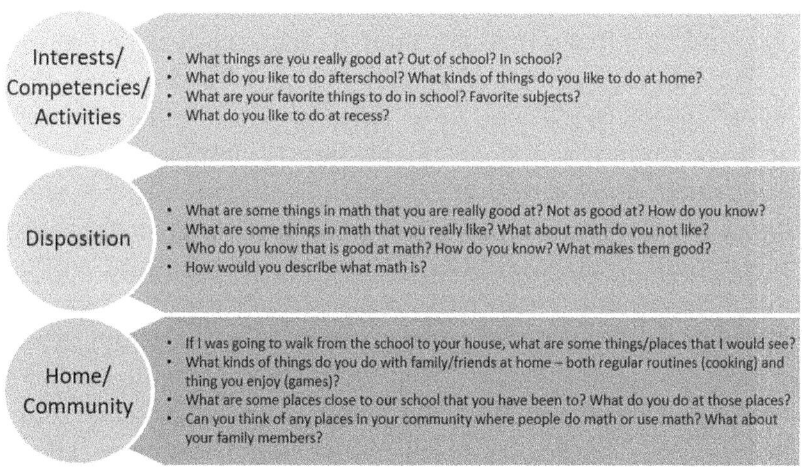

FIGURE 3.1 Questions for an informal interview
(summarized from Drake et al., 2015)

that you can use to get to know your learner's broader prior knowledge and experiences.

Informal Interview with Stu

When we spoke with Stu, we learned about his interests and his efficacy and identity toward mathematics in school, at home, and in the community. For instance, we learned that Stu was very excited and comfortable when working on any type of problem involving equal sharing, as he often needed to share with his younger sister at home. We also learned that Stu often held back in explaining his thinking during math class because he said, "The teacher is better at math", and often grew frustrated during math class because he did not consider it to be something he was good at.

Tasks and Task-Based Interviewing

In the beginning of this chapter, we mentioned that tasks act as platforms in which to discover and bring forward learners' prior knowledge specific to mathematics. To determine learners' prior mathematical thinking, tasks should be designed *wide* to ensure access to many ways of engaging in, representing, and expressing knowledge. These design considerations are often referred to in a framework called Universal Design for Learning (UDL) (see CAST, 2018). For the purposes of this chapter, we will think of *wide tasks* as those that allow for several possible solution strategies and they may even have more than one correct solution.

Let's unpack wide tasks a little bit more. What makes a task "wide" or not "wide" and how do we decide? For us, *wide tasks* are those that invite many possible means of engagement, representation, and expression – they allow several possible solution strategies. Using frameworks like UDL encourages teachers to intentionally design tasks so that they anticipate learner diversity and ensure learners access to their prior knowledge. Their lessons are designed to ensure learners access to their prior knowledge because they help us anticipate learner diversity and incorporate ways to invite and support it as a part of task design. To give you a brief introduction to the UDL framework, take a look at the questions in Table 3.2 that a teacher might ask themself as they intentionally plan with the UDL framework in mind.

TABLE 3.2 Planning framework for anticipating learner diversity in intervention planning (summarized from CAST, 2018)

Engagement

- How might you recruit learner interest?
- How might you take into account learner experience or inexperience on the topic?
- How might you vary the pace of work, length of work session, or timing and sequence of activities?
- How might you involve all participants in the whole class discussion?
- How might you encourage the creation of goals or task analysis of the goal into short-term goals?
- How might you provide frequent, timely, and specific feedback?

Representation

- How might you present or display information in varied ways?
- How might you clarify vocabulary?
- How might you activate or provide background knowledge?

Expression

- How might you vary your ways of responding?
- How might you use multiple media tools for communicating responses?
- How might you help learners set goals around this learning (feedback loop)?

To continue to support you in thinking about this idea, we invite you to explore the tasks in Table 3.3 below. Choose one and solve it; then solve it another way. As you work, consider which of the tasks below are "wide" for you and why?

All learners – including learners whom we describe as struggling with mathematics – use what they already know to understand and make sense of new ideas. It is the use of this existing knowledge that can pave the way toward new understandings. Understanding how learners think within wide tasks can give you information about their reasoning and help you use additional planning moves that are unpacked later on in this book (e.g., other kinds of tasks, teaching moves, or structures) and that can promote advancements in learners' reasoning.

To uncover a learner's mathematical prior knowledge, the teacher should start by having a variety of wide tasks that they believe to be within the learner's conception as well as some that

TABLE 3.3 Wide tasks – Why or why or not?

Grade band	*Task*	*Why this task is wide*
Elementary	Begin with any start number and add or multiply by the same amount. For example, start with 3 and add 5 or start with 4 and multiply by 2 . Complete the task *again* with a different start *or* change number (do not change both). Create a visual to record the pattern(s) you see.	• Allows for a variety of patterns, which can vary based on both the start number and the jump number. • Learners can select a variety of ways to show the pattern. They may want to make a list, hundreds chart, or table. They could use manipulatives of different colors that might help them see the patterns. • Since you repeat the task by changing *only* the start number *or* the jump number, the patterns you set up can be wide (allowing the learner to select both their start and jump number, or the teacher can select one).
Middle grades	Jessica can tile an 8-by-8-foot wall in ¾ of an hour. How long will it take her to tile another wall that is 7 by 8 feet?	• Allows for multiple representations in how the learner shows their thinking (pictures, double number line, table, or graph). • Allows for a variety of strategies to solve (multiplicative relationship, unit rate, other rates).
High school	How do you see the pattern growing? 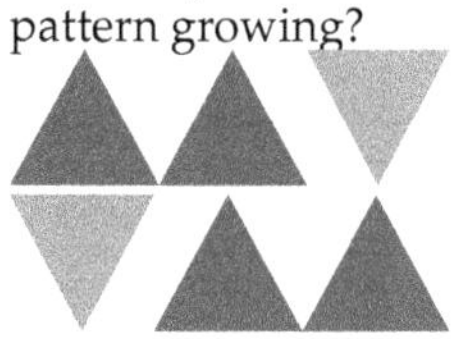	• Learners can select a variety of ways to show the pattern. They may want to make a picture continuing the pattern, table, graph, or equation. They could use manipulatives of different colors that might help them see the patterns.

may be bridging into new (not-yet-learned) conceptions. Possible ways for teachers to identify content for potential tasks include prior standard assessment information, grade-level standards, progressions documents, or curriculum tasks. At times, the intervention content that is being planned for learners to work within might be a concept or skill that is foundational to the intervention. It is often helpful to examine content that stretches across years to get a holistic sense of children's development of understanding over time.

Once you have the content focus, you can prepare clear and specific problem tasks to use. Tasks should be worthwhile to learners, placed in familiar contexts, and wide enough to elicit their thinking. This is where you can and should use what you found out about the learner during observation and the informal interview. Mathematical tools such as linking cubes, counters, or other tangible objects should be readily available for learners to choose from and use to show or support their reasoning. Finally, because the goal of presenting the tasks is to talk to learners and figure out how they think, teachers should have a few questions ready to ask learners regarding their work in the tasks. Questions like those shown in Figure 3.2 are often helpful starters and are all variations of the essential question "How did you solve it?" (Ginsburg, 1997). The learner's strategies, words, and drawings are windows into their understanding.

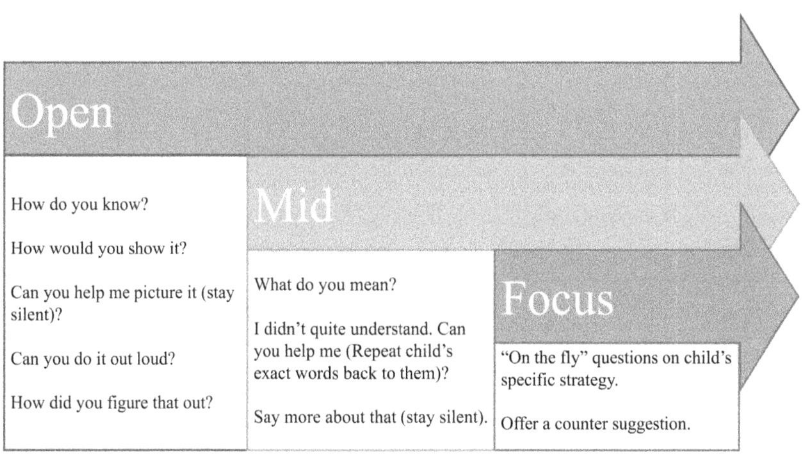

FIGURE 3.2 Questions to ask learners to uncover their reasoning in a task

Planning for Stu's Tasks and Task-based Interview

Because we knew that Stu's favorite thing to do at home was bake cookies with his mom, that he often shared those cookies with his sister, and that his favorite part of the school day was lunch, we created two tasks for Stu that were situated in the context of sharing cookies (baked by his mom) with friends from school during lunchtime. The context fit well with the content focus of partitive division with and without remainders and connected well to fractions. We then added other sharing situations that resulted in fractions by using some of Stu's favorite foods and names of his friends that he sat with at lunch. We also included other tasks that involved sharing but not food (i.e., sharing clay sticks) because we observed that Stu seemed to love his art class, as he often spoke up a lot during class, interacted with many learners, and shared ideas for drawings and sculptures.

To keep organized, we used a tool that looked much like Figure 3.3 when working with Stu since it allowed us to list our tasks and questions together and gave us a way to record what happened during the interview. The numbers in the brackets in the figure mean that teachers can allow for choice in number values.

<table>
<tr><th>Sample Problem Task</th><th>Questions to Ask (general)</th><th>Questions to Ask (specific)</th><th>Child's Representations/ Strategies</th><th>Child's Gestures/ Talk</th><th>Teacher Talk/ Interactions</th></tr>
<tr><td>Four friends shared [12; 13] cookies at lunch time. How many cookies did each friend receive?

Three friends share 5 slices of pizza equally. How much pizza does each friend get to eat?</td><td rowspan="3">How did you figure it out?

Can you help me picture it?

What if they want to [eat/share] all of it (if shares not exhausted)?

Say more about what you meant there.

They want to share this, too (if the shares are not exhausted).

What if they each want the same amount (if shares are uneven)?

I worked with a child earlier and they showed me this… what do you think?</td><td rowspan="3">How much did each person receive (what was their share)?

Can you count that out for me?

[If child quantifies as whole number] Four what? Four wholes?

You split those into [two, four] pieces. What do we call each piece?

Are halves and fourths the same thing/size? Why/why not?

How many [halves/fourths] fit back inside the whole?</td><td></td><td></td><td></td></tr>
<tr><td>Seven friends share four sticks of clay for an art project. How much of one stick of clay for each friend?</td><td></td><td></td><td></td></tr>
<tr><td>[4, 9] friends each use one-third of a container of paint to finish their project. How much paint is used?</td><td></td><td></td><td></td></tr>
</table>

FIGURE 3.3 Organizing tasks and questions

How to Introduce the Interview to the Learner

When tasks, tools, and questions are ready, it is time to talk with your learner. We suggest that you talk with learners individually, yet realize that this may not always be possible. In either case, teachers can begin by explaining to the learner what will happen in the interview and why it is important (Hunt, 2015). The best way to do this is to state, simply and honestly, the purpose for the interview. Getting any learner to show you their thinking, especially those who may not have a positive relationship with mathematics, might be difficult. For instance, learners might believe that it is not their role to show their thinking in mathematics. An example of how to begin these conversations that we have found to be effective as reported in Hunt (2015) is supplied in Figure 3.4.

During your conversation with the learner, it is best not to interject or lead the learner. In other words, it is okay to engage learners in productive struggle. However, if learners are reluctant to show their thinking or seem frustrated, consider asking him or her for help by using the prompts listed in Figure 3.3. Learners love to be asked for help and feel valued when you do that. It also helps them to show their way of thinking instead of what they think you want to hear. Be sure to encourage EFFORT

Today, we are going to work with several math riddles to help me better understand how you think about math. You may think some of the riddles are really easy, and you may think others are kind of tough. As you think about each one, you can use any of these tools [reference the paper rectangles, pencil, and paper] that you like or even your fingers!

There are a lot of different ways to think about each math riddle, and I am not looking for a "right" or "wrong" answer. I am just interested in how you solve them. If you are not sure how to solve one, just let me know. Sometimes I might ask you questions so that I can understand what you mean a little better. Are you ready to help me out?

FIGURE 3.4 A way to begin the interview

as opposed to "right answers" – and be specific with feedback as learners solve problems. Most importantly, acknowledge the learner's way of doing and try your best to understand what, from the learner's frame of reference, makes sense.

To summarize, teachers can use observation, informal interviews, and tasks to employ Key Teaching Move #1, which is to *Bring Forward Learners' Prior Knowledge*. These three tools work together to provide the richest possible picture of both a learner's current understanding and their learning potential. Uncovering areas where the learner does not already have an abstracted method for solving problems is where teachers can pose a task that learners find problematic and work to expand upon their reasoning. Chapter 4 will elaborate on this idea through the use of Key Teaching Move #2: *Promote Activity and the Need For Change*.

References

CAST (2018). *Universal design for learning guidelines version 2.2 [graphic organizer]*. Wakefield, MA: Author.

Drake, C., Aguirre, J. M., Bartell, T. G., Foote, M. Q., Roth McDuffie, A., & Turner, E. E. (2015). TeachMath Learning Modules for K-8 Mathematics Methods Courses. *Teachers Empowered to Advance Change in Mathematics Project*. Retrieved January 3, 2021 from www.teachmath.info.

Fuchs, D., Fuchs, L. S., & Vaughn, S. (2014). What is intensive instruction and why is it important? *Teaching Exceptional Children*, 46(4), 13–18.

Ginsburg, H.P. (1997). *Entering the Child's Mind: The Clinical Interview in Psychological Research and Practice*. Cambridge, UK: Cambridge University Press.

Hunt, J. (2015). How to Better Understand the Diverse Mathematical Thinking of Learners. *Australian Primary Mathematics Classroom*, 20(2), 15–21.

Hunt, J. H., Welch-Ptak, J. J., & Silva, J. M. (2016). Initial understandings of fraction concepts evidenced by students with mathematics learning disabilities and difficulties: A framework. *Learning Disability Quarterly*, 39(4), 213–225.

Institute for Learning (2015). *Multiplication of fractions.* Retrieved December 22, 2020, from http://ifl.pitt.edu.

Lo, J. J. (2004). Prospective elementary school teachers' solution strategies and reasoning for a missing value proportion task. *International Group for the Psychology of Mathematics Education*, 3(18), 265–272.

McLeskey, J., Barringer, M-D., Billingsley, B., Brownell, M., Jackson, D., Kennedy, M., Lewis, T., Maheady, L., Rodriguez, J., Scheeler, M. C., Winn, J., & Ziegler, D. (2017, January). *High-leverage Practices in Special Education.* Arlington, VA: Council for Exceptional Children & CEEDAR Center.

National Research Council, & Mathematics Learning Study Committee. (2001). *Adding it up: Helping children learn mathematics.* New York, NY: National Academies Press.

Nguyen, F. (2020). *Pattern #4, squares in step* 43 = 173. Retrieved December 22, 2020, from http://www.visualpatterns.org.

Van de Walle, J., Karp, K., and Bay-Williams, J. (2019). *Elementary and Middle School Mathematics: Teaching Developmentally.* Boston, MA: Pearson.

4

How Can Teaching Support Learners to Build Norms, Position Learners as Active, and Change Thinking?

Key Teaching Move #2: Promote Activity and the Need for Change

Action drives learning. If learners cannot actively think through a math problem, then they will have difficulty moving forward, both in the context of the content of the problem and in their mathematics reasoning more generally. As we shared in Chapter 2, we use the term "actively" to mean that learners have the opportunity to employ and reflect upon their own reasoning in a problem. However, for learners to use their own reasoning, they need to be supported to do so. In this way, among the most powerful tools that teachers have to support learning are the environment where learners engage and the activities that learners engage with.

The focus of this chapter is Key Teaching Move #2: Promote Activity and the Need for Change. We present the tools to promote learner activity and see the need for change through two lenses: (1) structuring Small Environments for success by co-constructing social norms and (2) selecting and sequencing tasks. We remind teachers that we define "Small Environments" both through the size of the intervention group and by how we position learners within intervention spaces and consider their potential for learning. Our job as teachers is to listen to learners' reasoning, figure out how they understand, and build on that understanding. Let's begin by thinking through structuring Small Environments for success by developing social norms.

Structuring Small Environments for Success: Developing Social Norms

Actively involving learners in developing and refining/redefining the norms within the Small Environment sets the stage for their active role in learning. Norms are a "big deal" – they are linked to learners' mathematical dispositions (NRC, 2001) and autonomy within group settings and are often thought of as social correlates to learners' evolving inner conceptions and beliefs (Cobb & Yackel, 1996). Supporting learners' dispositions and autonomy helps them to use their own knowledge in the Small Environment and advance their knowledge over time.

With respect to mathematics, norms can take two forms: (1) social norms and (2) sociomathematical norms (Cobb & Yackel, 1996). This chapter addresses social norms, which are regularities that have been normalized as the expectation of the Small Environment and are jointly constructed by both the learners and the teacher. They come from a place of understanding how learners and teachers best interact in the mathematics classroom environment. They include but are not limited to the following:

- Respecting and honoring the thinking of classmates and teachers
- Listening actively
- Attempting and engaging in work
- Offering respectful feedback.

When we think about the specifics of the mathematics classroom, we might add explaining and justifying solutions, an active attempt to make sense of explanations given by other learners, and agreeing and disagreeing on a particular solution. Social norms are not the same thing as learners' inner cognition or their thinking. Yet, both are complementary to learners' individual beliefs about mathematics as it occurs in school. Social norms and beliefs influence each other and support the participation structure in the Small Environment.

Social norms that learners currently hold with respect to mathematics learning environments can be surprising. Teachers often overestimate their learners' propensities to publicly explain and justify their own reasoning in whole class mathematics settings (Cobb et al., 1989). As opposed to accepting, using, and communicating their own reasoning as the norm, learners seemed more apt to attempt to infer what they felt the teacher was looking for. This view of norms is likely backed in intervention settings. For example, Jackson and Neel (2006) found that, on average, the time devoted to algorithm instruction for learners in special education classes was more than double the time devoted to algorithm instruction for learners in general education classes (about 30% in general education versus 75% in special education classes). The findings suggest limitations in terms of the active ways within traditional intervention environments that learners are able to learn.

In fact, contemporary research shows that norms held by learners with disabilities or learners who have participated in more traditional forms of intervention were in line with those described by Cobb et al. in 1989. In one study on learners' initial ideas about fractions as quantities (Hunt & Empson, 2015), learners often looked to the teacher for quick answers, modeling of solutions, representations, or other ideas about how to solve the problems and were quite hesitant to engage in their own reasoning. All of the learners involved in that study had taken part in supplemental mathematics interventions the previous two years, and their views of intervention time were largely in line with remediation settings mentioned in Chapter 1. For these learners, the job of the teacher was to generate mathematical thinking. The learners' job was to re-state the teacher's ideas as opposed to using and growing their own reasoning. In short, the norms developed during these learners' math intervention time supported passive views as mathematics learners as opposed to that of active thinkers and doers.

Passive learning norms have been found in other studies of learners with disabilities. Lambert et al. (2020) found that, compared with his peers, a learner with autism participated in mathematics discussions very infrequently. This learner did not seem to understand the learner-centered social norms at work in his mathematics classroom. Although this learner constructed viable solutions on his own, not understanding how to participate in the learner-based social norms was detrimental to his participation in classroom discussion. Finally, Hunt et al. (2019) found that a learner with specific forms of learning disabilities used the explicated thinking taught to her by her teachers when engaging in mathematical problem-solving. A more severe example of inferring what the teacher was looking for during discussions, this learner actually took on previously taught strategies and algorithms as her own reasoning. Taking up someone else's reasoning can be okay if that reasoning is understood. However, the norm proved detrimental to the learner's conceptual development in this case because although the child could recite the strategy, she could not explain how or why it worked or when it was appropriate to use (Hunt et al., 2019).

Getting Learners to Establish and Participate in Social Norms

Teachers must put in the work to explicitly change the expectation of who does the thinking during intervention time. The work begins by inviting learners to do mathematics. As we hinted at in Chapter 2, "we believe that if learners are to understand mathematics for themselves [though most certainly not by themselves], it is more helpful for thinking of understanding as something that results from solving problems, rather than something we can teach directly" (Making Sense, 1997, p. 25). Envision a Small Environment where learners are doing mathematics. What action verbs would learners use to describe what they are doing? Then check out Figure 4.1 and see whether any of the words listed in the figure match your list.

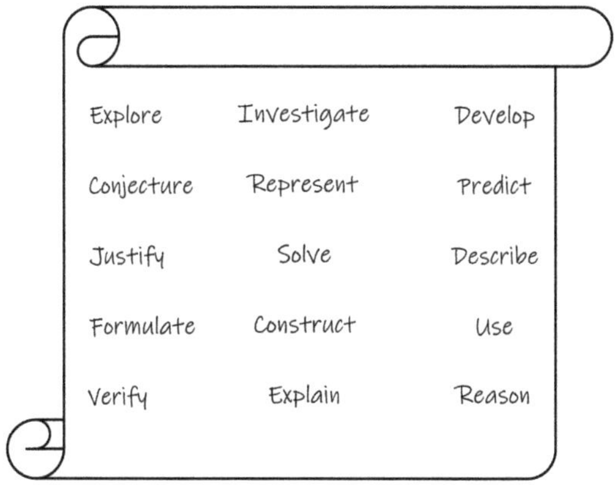

FIGURE 4.1 Learners in active Small Environments

Several of the words on the list can also be used to describe how teachers can support learners to work with each other to build social norms in the Small Environment. For instance, learners can be supported to develop or formulate norms, describe and use them, explore when norms work and when they need revision, and investigate new ones. Consider how this is different from simply supplying norms to learners and "enforcing" their use. Teachers might ask learners to do things like share ideas and opinions, ask and answer questions, invite quieter group members to speak, and so on. Yet learners might not take up such norms as more than a set of classroom rules, and not all learners will understand the norms if they did not have a hand in generating them.

The work of the teacher, then, is the facilitation of learners' co-construction of social norms in the Small Environment. One way to co-construct social norms is to actively involve learners in deciding what the norms for the intervention group will be and get learners to commit to and use the shared norms over time. Doing so can be helpful to counteract and change how learners view mathematics interventions – and their own activity as participants within them – over time.

Devote Lesson Time to Group Construction of Norms

One practical way to do this is to hold a class activity around what it means to engage in a mathematical discussion or what it means to share, explain, and justify mathematical thinking. Begin by leading a discussion with the learners in the Small Environment with a goal of establishing why it was important to engage in mathematical discussion. Teachers may do this by explaining that learners will each engage in problem tasks by themselves yet will transition to sharing their thinking as a group. It might help learners to "buy in" to the idea of sharing their ideas if you explain that this process will help them to make the necessary mathematical connections to help move their learning forward.

Next, you are going to want to create an anchor chart that can be prominently displayed and often referenced. On the left, write the words "looks like"; on the right, the words "sounds like." Building from the discussion around the importance and purpose of mathematical discourse and engagement, ask the learners what it "sounds like" and "looks like" to be engaged in mathematics. As learners generate ideas, document their responses on the chart paper. Once a good number of ideas are recorded, ask the learners which ideas seem similar and which are distinct. This step will ensure that learners make connections between similar ideas and will also shorten the list ideas to a core set of ideas and values of the group. Work together with your learners to refine the connected ideas into normative statements and compose a list; we've included a finished example below in Figure 4.2.

Finally, check in with learners regarding the final list. Ensure that each member feels heard and that the stated norms represent the original ideas generated by the group earlier in the process. Teachers might consider creating specific sentence starters (see Figure 4.3), inspired by the norms created and confirmed by learners on a clean sheet of chart paper. The chart can be posted at the front of the class. It is important to note that co-constructed and agreed-upon norms can and should be revisited or revised over time as needed.

If teachers engage learners in this simple yet powerful co-construction of norms, then learners are positioned as active members of the Small Environment *from the start*. This process can take an entire intervention session (or multiple sessions) to complete, yet the benefits to learners' development are well worth the time and the work! Putting the effort into social norms can help put learners in the driver's seat of their learning and development,

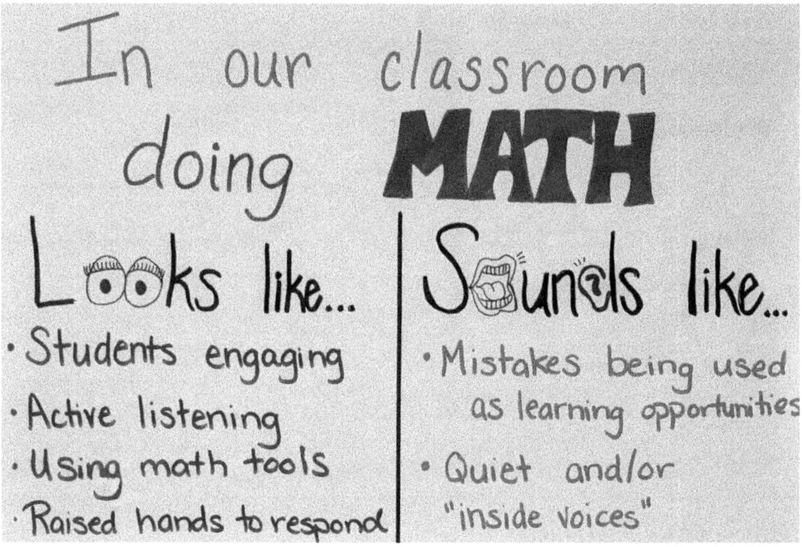

FIGURE 4.2 Social norms list

Active listening & responding	Sharing your ideas
"I agree with you because…."	"I have an idea to share…"
"I respectfully disagree with you because…"	"I solved the problem by…"
"I like how you… Have you thought about…"	"I wonder if…"
"I understand how you…"	"I noticed that…"
"Can you explain how you…"	"First, I … Then I …"
"How did you know…"	

FIGURE 4.3 Sentence stems/frames for developing from norms

influence learners' sense of mathematical competence, and set a foundation for learner autonomy in the Small Environment.

Selecting and Sequencing Tasks

Selecting and sequencing tasks are just as important as co-constructing social norms. In this section, we offer four classifications of tasks: ***Accessing***, ***Bridging***, ***Challenging***, and ***Varying***.

Task Type	***Description***	***Example*** *(using "Ed's" case from Chapter 1)*
Accessing *3-4 sessions*	A "wide" task that has many entry points such that learners can solve in many ways and that the teacher can deduce learner thinking	Two learners share five tacos equally. How much taco does each person get?
Bridging *1-2 sessions*	A task that is accessible from how a learner currently understands. Brings forward a learner's current way(s) of reasoning, or their prior knowledge	Two learners share one Taco equally. How much of one taco does each learner get? *No constraints*
Challenging *Varies; 2-4 sessions*	A task to support the same mathematical goal yet changes in terms of constraint(s) on solution method to challenge the learner's thinking	[Three; Four] friends share one french fry equally. How much does each person get? *Add **Constraints**: No folding or marking the fry.* [Five; Six] friends share one french fry equally. How much does each person get? *Constraint: Find the correct length in as few attempts as possible.*
Varying *Varies; 2-4 sessions*	Different tasks in which the same solution method can be applied across them	Here is [one-third; one-tenth; one-seventh] of a whole french fry. Show the length of one whole french fry. *Parts should iterate to different whole lengths in terms of the referent whole.* We have one-third of a whole french fry and one-fifth of the same whole french fry. Which part of the fry is longer? Why? *Parts should not be visible to learner*

FIGURE 4.4 Task sequencing

For a preview of task classifications, see Figure 4.4. We offer these classifications as a framework that teachers can use to think about and sequence tasks for instruction in the Small Environment. Let's explore each one below.

Accessing Tasks

Accessing tasks allow two things: (1) a way to access, use, and explore the learners' own reasoning and (2) a way for the teacher to observe the learners' thinking without interruption. Accessing tasks are purposely ***wide*** because they allow for many pathways of reasoning through them. They also tend to have long lives, so to speak, because the mathematics that is supported by them spans time or even grade bands.

Thinking about Ed's Thinking

Consider the case of Ed from Chapter 1. Because we did not yet know Ed's thinking, we posed the equal-sharing problem of sharing five items among two people as our first Accessing task. The equal-sharing problem posed to Ed is wide because it can be accessed through whole number notions of grouping and partitive division. Second- or third-grade learners can engage in these tasks! Yet it "lives" long into middle-grade notions of distributive reasoning and even proportionality (Empson & Levi, 2011; Silva et al., revisions).

Ed's initial thinking gave the teacher a view of his prior knowledge. From our asset-based and process-oriented frame of reference, we can tell from Ed's thinking that he knows about equality because he deals out the tacos, one by one, to the sharers. We also know that Ed knows that he did not have enough to make equal shares in terms of whole tacos since he initially pushes the final taco to the side as he considers his actions in the problem. In fact, Ed's initial solution was to build out groups of two, so he has ideas about two as a whole number group or unit that he could use to build upon in terms of fractional units.

Possible Pacing

As tasks are designed for the Small Environment, it is a good idea to begin with several "Accessing" tasks, especially if teachers have not had a chance to engage in any form of learner-based mathematics work to determine learners' prior knowledge. (See Chapter 3 for examples and Chapter 6 for tools to document learner thinking during Accessing tasks.) That is, if teachers find themselves beginning instruction without a good handle on the prior knowledge of their learners, then beginning a task sequence with at least three sessions' worth of Accessing tasks is standard in terms of task selection and sequencing. Only after observing your learners in Accessing tasks (through a revolutionized and process-based lens, of course!) should you continue interventions with the next task in the sequence: Bridging tasks.

Bridging Tasks

In **Bridging tasks**, learners solve tasks that activate their available concepts. That is, Bridging tasks act as platforms for learners to access what they already know so they can make connections, or *bridge*, their prior knowledge to a new idea. For example, we chose to ask Ed to equally share ONE item between two people. The item could be a paper strip that represents something familiar to the learner, like a French fry (Tzur & Hunt, 2015). Can you think about why this particular task is a Bridging task for Ed? Our justification is on our notepad shown in Figure 4.5.

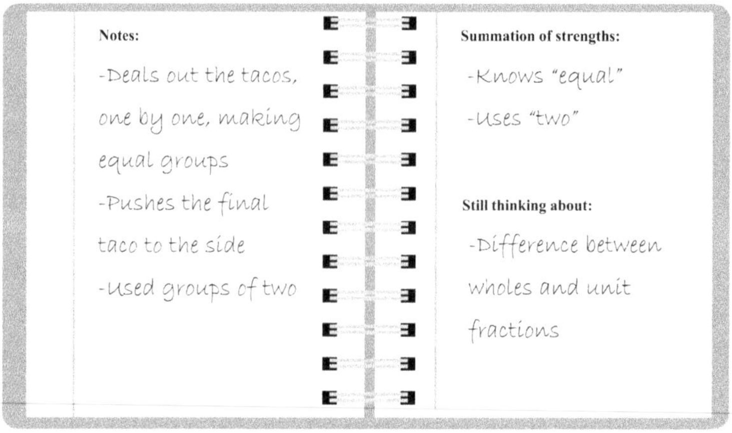

FIGURE 4.5 Notation of strengths

This is likely a bridging situation for Ed because he has already used a concept for two that he brought into a problem and used to think about "equal shares." He has also attempted to keep equal shares and use up a referent quantity (in this case, the five tacos). He was also bothered when there were "leftovers" in his solution.

The new problem we propose – equally sharing ONE item between two people – acts as a bridge to new knowledge from Ed's prior knowledge because it invites him to (a) consider keeping equal shares and (b) exhaust a whole quantity that now (c) results in a fraction and not a whole number. The Accessing problem also invited Ed to think about (a) and (b), yet the presence of many items to share did not invite Ed to focus on one item to create fraction quantities (c). The task of equally sharing ONE item between two people may **not** be a bridging situation to Ed or other learners if they have different prior knowledge (e.g., Hunt, Silva & Lambert, 2019).

The task may also not be bridging for Ed if we do not fully consider his prior knowledge (see Chapter 3). For example, we used tools such as those described in Chapter 3 to know that Ed shared French fries at home with his siblings. Bridging tasks should bring forward *all* of a learner's prior funds of knowledge so that they can use all of themselves to understand. Our example above shows all of the mathematical prior knowledge that Ed used in our Bridging task to understand. Yet it is important for teachers to note that learners have other funds of knowledge that are based in their experiences and in their cultures and that they can also use to understand. Get to know your learners and be sure that the contexts that you use in tasks make sense to them. Using all of a learner's prior knowledge is important because if we do not, then we risk blocking access to the bridge, so to speak. That is, if the context of the task conflicts with the knowledge that a learner brings in, then it will not bridge to the mathematics as we intend.

Thinking about Ed's Potential Thinking

Let's think about how Ed might work in our Bridging task with a context that made sense to him. Ed might bring forward his prior knowledge to consider how to share the fry among two people and not have any leftovers. For instance, he may make two parts by cutting or folding the paper and notice they are very different in size. Without any constraints on the task (i.e., asking him not to fold the paper or asking him to use one partition to show the first share only), he may cut the fry apart and directly compare the parts. Because the parts are not the same length, Ed could use another part of his prior knowledge – the notion of "half." This notion helps Ed to fold the fry into two parts before he cuts it apart. Now, the parts are equal in length AND use the entire length of the whole fry.

Possible Pacing

The issue of pacing in Small Environments is tentative, as our goal is to respond to learner thinking. This is to say that the number of sessions in which you would use Bridging tasks will change based on learners' thinking. Yet teachers may generally plan for one to two sessions to use a Bridging task in the Small Environments. Chapter 6 will expand on the notion of pacing and how you can pair data on learner thinking with the task selection and sequencing framework to make planning and instructional decisions.

Challenge Tasks

Challenge tasks promote the need for change by challenging learners' current thinking so that it begins to expand. Challenge tasks engage learners in solving gradually more difficult tasks by using one type of task in different ways. For example, in the case of Ed, a Challenge task may be to share the same-size paper strip of paper (representing a fry in the Bridging task) among *three* people, including a **constraint** that he is no longer allowed to fold the fry or tear apart created pieces to visually compare results.

When we say that Challenge tasks are the same type of task yet are solved in different ways, we mean that the tasks support the same math goal yet change in terms of constraint(s) on the solution method. For example, additional tasks can engage Ed to equally share the same-size paper strip among 4, 5, or even 10 people. Along the way, we can impose constraints (e.g., find the correct share length in as few attempts as possible and do not fold the paper). The goal of the tasks is consistent – to create equal parts by repeating an estimate of one share length and coordinate the equal shares within the length of the whole paper strip. The constraints promote the need for more and more sophisticated reasoning that connects to how Ed is already reasoning in the task.

Thinking about Ed's Potential Thinking

Within the task, Ed may bring forward his concept of three, marking three parts within the fry and initially attempting to directly compare the three parts as he did in the previous situation. Yet, this time, we pose a constraint in the task: we ask that Ed not fold the fry or cut it apart until he is sure that he has created same-sized pieces. With the constraint, Ed uses the length between his thumb and pointer finger and repeats it across the fry (i.e., the repeat strategy, Tzur, 2007; Tzur & Hunt, 2015) to compare their lengths through indirect measurement. Seeing that they are not the same length, Ed creates a new strategy: he makes an estimate of the length of one share with his fingers, repeats it to make three shares, and compares the length of the three shares to the given whole. The use of the constraint of finding the right share length in as few attempts as possible supports Ed to become more and more precise in his estimates.

Possible Pacing

As stated above, the number of sessions in which you would use Challenge tasks will change based on learners' thinking. In our experience, up to four sessions use Challenge tasks in the Small Environments. See Chapter 6 for how you can pair Challenge tasks with data on learner thinking and know when to infuse Challenge tasks into the sequencing framework to maximize results for your learners.

Varying Tasks

Varying tasks are related to Challenge tasks yet invite learners to solve a *different task* (or tasks) using the same method used in the Challenge task. Varying tasks should naturally connect to the content involved in the Challenge tasks (e.g., in Ed's case, using a created unit fraction to reform the length of a whole; "Here is one third of a whole French fry. Show the length of one whole fry"). In this way, Varying tasks foster learners' noticing of what stays the same across situations and why that "sameness" is mathematically valid and justified (Simon et al., 2004). In other words, the Varying tasks invite learners to use thinking formed in the Challenge tasks in a new way and invite that thinking to grow.

Thinking about Ed's Potential Thinking

In Ed's case, the strategy he was using – the repeat strategy – is also used in the Varying tasks but with a different goal. Namely, Ed can use the strategy in a connected task of using a unit fraction to create the length of a whole. This task serves as a platform for Ed to connect the creation of unit fractions from a whole to the creation of a whole from a unit fraction. These actions connect to a learning goal: there is a single unit that uniquely fits n times within a given whole.

Possible Pacing

As stated above, the number of sessions in which you would use Varying tasks will change based on learners' thinking. In our experience, up to four sessions use Varying tasks in the Small Environments. See Chapter 6 for how you can pair Varying tasks with data on learner thinking and know how and when to integrate Varying tasks with Challenge tasks to maximize results for your learners.

Earlier in the chapter, we proposed that an intentional focus on co-construction of classroom social norms is a useful and necessary way that teachers can position learners as thinkers and doers of mathematics in Small Environments. We also looked at how to select and sequence the four classifications of tasks to ensure that learners are advancing their learning through a systematic and achievable process. As we look to Chapter 5, we will unpack Key Teaching Move #3: Promote Noticing, Reflection, and Connections. A connected aspect of promoting learners as active members of the group is through the use of routines to implement mathematics problems or tasks. We offer a common structure for each task used in Small Environment instructional sessions: the **Think-Pair-Share**.

References

Cobb, P., & Yackel, E. (1996). Constructivist, emergent, and sociocultural perspectives in the context of developmental research. *Educational psychologist*, 31(3–4), 175–190.

Cobb, P., Yackel, E., & Wood, T. (1989). Young children's emotional acts while engaged in mathematical problem solving. In P. Cobb, E. Yackel, & T. Wood (Eds.) *Affect and Mathematical Problem Solving* (pp. 117–148). New York, NY: Springer.

Empson, S. B., & Levi, L. (2011). *Extending Children's Mathematics: Fractions and Decimals*. New York, NY: Heinemann.

Hiebert, J. (1997). *Making sense: Teaching and learning mathematics with understanding*. Portsmouth, NH: Heinemann.

Hunt, J. H., & Empson, S. B. (2015). Exploratory study of informal strategies for equal sharing problems of students with learning disabilities. *Learning Disability Quarterly*, 38(4), 208–220.

Hunt, J. H., Silva, J., & Lambert, R. (2019). Empowering students with specific learning disabilities: Jim's concept of unit fraction. *The Journal of Mathematical Behavior*, 56: 100738.

Jackson, H. G., & Neel, R. S. (2006). Observing mathematics: Do students with EBD have access to standards-based mathematics instruction?. *Education and Treatment of Children*, 593–614.

Lambert, R., Sugita, T., Yeh, C., Hunt, J. H., & Brophy, S. (2020). Documenting increased participation of a student with autism in the standards for mathematical practice. *Journal of Educational Psychology*, 112(3): 494.

National Research Council, & Mathematics Learning Study Committee. (2001). *Adding It Up: Helping Children Learn Mathematics*. New York, NY: National Academies Press.

Simon, M. A., Tzur, R., Heinz, K., & Kinzel, M. (2004). Explicating a mechanism for conceptual learning: Elaborating the construct of reflective abstraction. *Journal for Research in Mathematics Education*, 305–329.

Tzur, R. (2007). Fine grain assessment of students' mathematical understanding: Participatory and anticipatory stages in learning a new mathematical conception. *Educational Studies in Mathematics*, 66(3), 273–291.

Tzur, R., & Hunt, J. (2015). Iteration: Unit fraction knowledge and the French fry tasks. *Teaching Children Mathematics*, 22(3): 148–157.

5

How Can Teaching Be Used to Promote Learners to Reflect Upon and Discuss Their Thinking?

Key Teaching Move #3: Promote Noticing, Reflection, and Connections/Conversation

The action that learners use to solve tasks that we discussed in Chapter 4 must be combined with noticing, reflecting, and conversation of mathematical ideas in order to move the learning forward. When learners notice what happens as a result of their actions as they work in and across relevant situations, they begin to advance their learning. When learners repeatedly reflect on connected and distal experience, they can solidify learning. Repeated reflection can be accomplished by reflecting on one's own actions or on others' actions through conversation. Together, noticing, reflecting, and communicating support learners' anticipation of mathematical ideas over time. But how can teaching be used to promote learners to notice, reflect upon, and discuss their thinking?

As we shared in the last chapter, we will offer a common structure in this chapter to facilitate not only employment of tasks but also support for learner reflection, noticing, and conversation in Small Environment instructional sessions: the

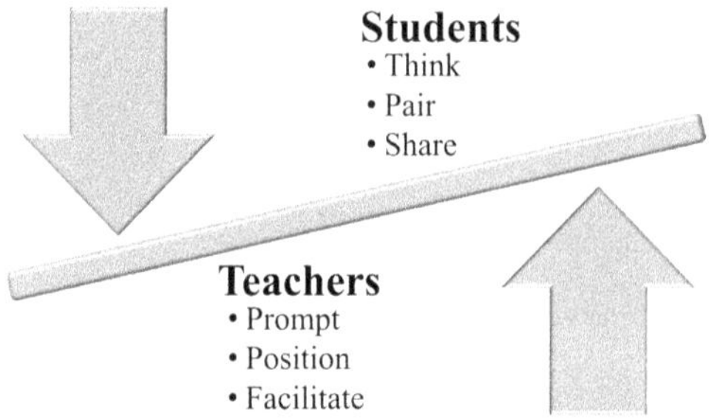

FIGURE 5.1 Learner and teacher roles during a Think-Pair-Share

Think-Pair-Share. In Figure 5.1, we preview the use of the structure from the lens of the learner and the lens of the teacher. In the following paragraphs, we will highlight the learner's work in a Think-Pair-Share and how teaching is used to support learners' mathematics.

Basics

The basic intervention structure that we describe in this chapter is Think, Pair, and Share. As you probably already know, in a Think-Pair-Share, learners work on tasks first on their own, then in pairs, and then as a whole group (a "share" as named by the teacher). Although this structure may seem straightforward, we highly recommend that you model each part of the structure with your learners so that they understand how each part is used and how the parts fit together to structure intervention time. Having a successful implementation of the structure will support learners to come to value and look forward to each of the parts. However, if it doesn't go well initially, be honest with your learners about what went well and what needs to be adjusted for the next time.

In the following paragraphs, we will explain the teacher's role in the use of the structure to support learners during "Think", "Pair", and "Share" for each task posed in Small Environments. Within each section, we will unpack the moves of the teacher to prompt, position, and facilitate learner success.

Learners "Think", Teachers Prompt

Learners Think

"Think" time is the first part of the Think-Pair-Share structure. It gives learners a quiet space to generate, notice, and reflect upon their *own* thinking in relation to the problem. Making sure that you are providing learners with an adequate amount of time is critical, as this is often an area where time is cut short. During this time, give learners a wide task to solve (see Chapter 3). Encourage them to record their thinking in whatever form makes sense to them. Have a variety of tools available for learners to use as they work. That is, think about including tangible (e.g., linking cubes and Cuisenaire rods), semi-concrete (e.g., ways for learners to draw out their thinking, such as white boards and markers), or even virtual tools for learners to think with. Choose tools that provide a variety of ways for learners to make sense and also align with the mathematics supported in the problem.

Teachers Prompt

Tasks can act as platforms for learners to think within. Yet research has found that learners who have difficulties learning mathematics benefit from thoughtfully planned interactions with the teacher as they think (e.g., Hunt & Silva, 2020; Hunt & Tzur, 2017). These interactions can help learners to *notice* and *reflect upon* their reasoning so that connections are made explicit. Noticing and reflecting are connected to self-regulated learning with respect to goal setting, monitoring, and reflecting during problem-solving. In Table 5.1, we list general questions and actions that you may consider to present to learners as they notice, reflect upon, and plan their thinking.

Let's think about how to design prompts to support learners. Generally, prompts are always attached to strategies that you anticipate the learners will use in a problem task. Get in the habit of anticipating at least three strategies that your learners may use to solve a task, so that you can create prompts to respond to learner thinking. In this book, we have supplied the anticipated learner thinking for you to create prompts for. Look at the table below and the learner work for the task, *"Three learners share two*

TABLE 5.1 Prompts and gestures to support learners during "Think" time

Prompts and Gestures to Support Noticing	
Prompt: Re-voice	You said.... Is that what you said?
Prompt: Cause and Effect	What happened when you...? Is that what you thought would happen?
Gesture: Re-show	*Re-show the learner's exact actions in a task.*
Prompts and Gestures to Support Reflection	
Gesture: Re-show	*Re-show the learner's exact actions in a task.*
Prompt: Justify	Convince me how you knew that... *(directly following a cause-and-effect situation)*
Prompt: Make Prediction or Apply Reasoning	How will you...? Tell me first, before you do it. Can you use what you did before to help you here? Is there a way for you to... [specific to connecting problem-solving strategies]?

granola bars." While the learner was working on this task, they had drawing materials, paper strips, and linking cubes available to them to support them in their thinking. Examine the learner work in Figure 5.2 and make observations based on the questions in Table 5.2. Ask yourself, "What might you ask to support this learner to notice their own thinking, reflect upon it, and plan their next problem-solving strategy?"

Below, we show the table again, yet this time it is filled in with what we would suggest asking this learner (Table 5.3). First, the learner either makes unequal shares or uses half to partition yet leaves the strategy incomplete. From these actions, we want to support this learner to *notice* what happened when they partitioned in the manner they did and the result of the partitioning. Next, we want to support the learner to *reflect* upon whether their action in the problem led them toward or away from the goal of making equal shares for three people. Finally, we want to support the learner to reflect upon how they may partition the items given another opportunity.

To summarize, research has found that learners who have difficulties learning mathematics benefit from thoughtfully planned interactions with the teacher as they work through problems.

Task: Three learners shared two granola bars

- Learner initially took both strips of paper and folded them in half and drew the red line on the top one. He counted "1, 2, 3" and paused. He put the strips to the side.
- Next, (top) the learner drew the two strips and cut the top strip in half. The faces represent each person. He counted "1, 2, 3" and paused.
- Next, he said, "I don't know," and then drew the number sentences (bottom).
- At this point, the teacher prompted by asking, "How would you share one bar among the three?" The learner drew on rectangle, followed by the third. The learner counted "1, 2, 3" and did not attempt to name the drawing.

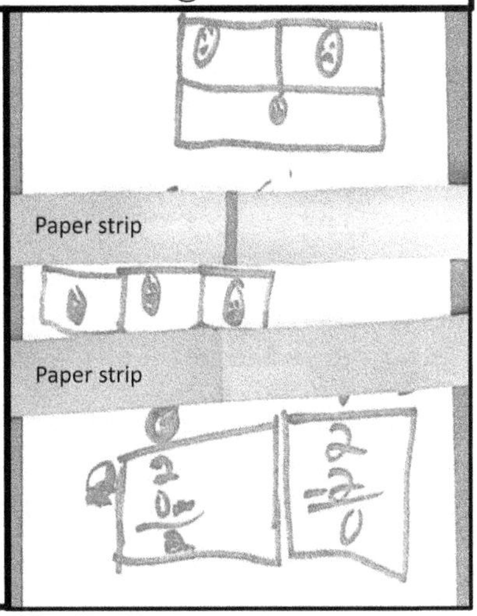

FIGURE 5.2 Learner thinking

TABLE 5.2 Planning questions to use during "Think" time

Question	*Response*
What are your observations about this learner's work?	
What might you say to help the learner **notice** their thinking?	
What might you say to help the learner **reflect** on their thinking?	

These interactions support the learner's noticing and reflection of their own actions in a problem. Prompts also support learners to use their previous thinking in a forward way so that connections are made both within and across different problem situations. Writing prompts is a generative practice that teachers can use alongside anticipating learner thinking to support learners as they "think" through a task.

Learners "Pair", Teachers Position

Learners Pair

During "Pair" time, learners pair up in order to share their thinking with a partner. Although the pairing of learners can

TABLE 5.3 Planning questions to use during "Think" time (completed)

Question	*Response*
What are your observations about this learner's work?	• *Halves the items or leaves whole (paper items; upper right)* • *Draws parts and counts "1, 2, 3" as they connect the parts.* • *Does not attempt to name the equal share.*
What might you say to help the learner **notice** their thinking?	*"What happened when you cut the bars into two parts? Is that what you thought would happen?"*
What might you say to help the learner **reflect** on their thinking?	*"Did your plan for sharing use up all of the bars and give everyone an equal share? Why/why not?"* *"What if there was only one bar to share?"* *"Is there a way for you to cut the bars so that you use up all of the bars and give everyone an equal share?"*

be random, we strongly encourage teachers to be intentional because this part of the structure supports the collaborative culture that you are setting up in your classroom. It increases depth of learner understanding and might give learners further opportunities to adjust or extend their thinking. This should not just be a time where one learner shares while the other listens, followed by the second learner sharing while the other listens. This should be a time where learners are listening for understanding as well as an opportunity for them to refine their thinking on the basis of their conversations. But how can you support learners to talk to each other during this time? Some learners may not think it is their role to discuss their mathematical thinking.

Teachers Position

The teacher's role during "Pair" time is critical because learner conversation during this time sets them up for success to "Share" their thinking in the larger intervention group. One way to begin is to use sentence stems or frames, such as those shown in Figure 5.3, to get the conversation started. Sentence stems and frames are beginnings of sentences or portions of sentences that you can place on display for learners to use to begin to talk to

Active listening & responding	Sharing your ideas
"I agree with you because…."	"I have an idea to share…"
"I respectfully disagree with you because…"	"I solved the problem by…"
"I like how you… Have you thought about…"	"I wonder if…"
"I understand how you…"	"I noticed that…"
"Can you explain how you…"	"First, I … Then I …"
"How did you know…"	

FIGURE 5.3 Sentence stems/frames for "pair" time in intervention lessons

others about their thinking. Having these available for learners to use assists learners to put a voice to the thinking shown on paper or with objects.

Sometimes, sentence stems on their own are not enough to get learners talking. There are additional things that you as the teacher can do to support learners to share their thinking. First, consider the physical setup of "Pair" time. Invite learners to share as a pair by encouraging them to push their papers together to invite conversation. Or call for a "heads together" – a common collaborative grouping structure. Consider what norms need to be developed in order to best support an environment where learners have ownership in collaborating with their peers. Might it help to have a norm where learners are allowed to get up from their seats to ask a question of another learner? We will revisit norms toward the end of this chapter. Finally, invite learners who have not yet spoken by saying, "You look like you have something to say. What is it?" Or "You look like you have an idea. What is it?" These small touches position learners as competent, invite them into conversation, and make them feel at ease and comfortable to share their reasoning with each other.

Second, teach learners to listen to each other with understanding. Listening for understanding can be a difficult thing to model. One simple way to do this is to remember the phrase "Yes, and…". Though simplistic, saying "Yes, and" (as opposed to "No" or "Yes, but") is a quick way to model how learners can to listen to each other as strategies and solutions are being shared.

TABLE 5.4 "Yes, and" versus "Yes, but" and "No"

Student A says	"I folded each paper strip into two parts, but I had an extra part."		
Student B response	"**No**, that's not what I did."	"**Yes, but** that's not how I thought about it."	"**Yes, and** I folded the paper strips into three parts."
Student A response	Shrinks and shuts down.	Gets quiet. Doesn't say anything else.	"Why did you fold it into three parts?"

Consider the three examples in Table 5.4. Notice the impact that student B's response has on the conversation in each case.

Modeling the phrase "Yes, and…" in natural ways, such as circling the room during "Pair" time and interacting with pairs as they share their thinking, is often best. This time should also be used to monitor the learners' conversations and the variety of learner thinking. You may want to consider which solutions you would like to discuss during the "Share" portion of the structure and in what order.

Learners "Share", Teachers Facilitate

Learners "Share"

The share part of the intervention structure is arguably the most important part of an intervention lesson. This part is where learners make intentional connections to *each other's* thinking and to the big idea of the lesson.

Teachers Facilitate

Conversation should be structured by the teacher to ensure that discourse takes place among the group. We give a few recommendations that you may consider regarding how to choose work to be discussed during "Share", how you may sequence who shares, and talk moves that will encourage meaningful group discourse.

Selecting and Sequencing Learners' Thinking

Once learners have engaged in "Think" and "Pair" and generated a strategy, how might you decide whose thinking will be shared or who will share in what order? Teachers often use choice and let learners volunteer to share. The benefit of this approach is that it is low-risk on the part of the learner – learners who want to talk can do so and those who want to observe can do so. The drawback of this approach is that you may have the same learners sharing all the time, unintentionally elevating some learners while marginalizing other learners' approaches.

Suppose you have the three learners in your intervention group as shown in Figure 5.4. They have just solved the problem, *"Five learners each eat ⅓ of a churro during a field trip. How many churros were eaten during the field trip?"* How would you choose to sequence learners' solutions? One choice may be to sequence responses beginning with the learner who demonstrates the most "correct" way of reasoning. For instance, you may begin with learner 1's thinking followed by learner 2 and finally learner 3. This choice is certainly plausible, yet using it all of the time may work to position some learners' thinking as more valued than others.

Working to find what is mathematical about *all* solutions and how that mathematics relates to the big idea of the lesson is the goal of sequencing (Smith & Stein, 2011). So it may be that you begin with learner 2, as this learner had an "oops" moment when he realized the iteration of one third five times would not

Student 1 **Student 2** **Student 3**

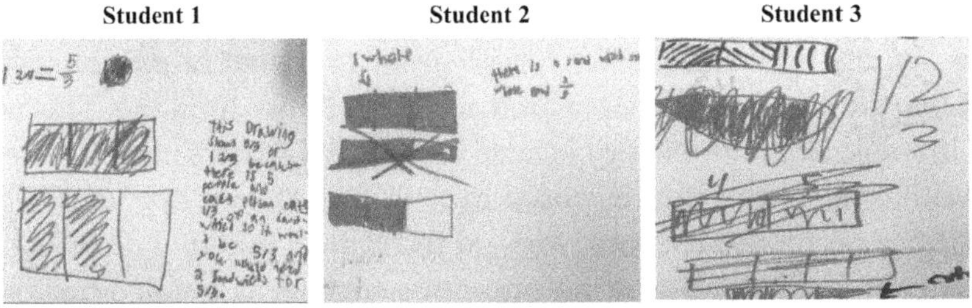

Student 1

- Verbally said "five-thirds" before drawing anything
- Drew two bars; explained they should be the same size and remarked his drawing was "a little off"
- Colored the first bar before partitioning it into 3 parts
- Partitioned the second bar into 3 parts, shades 2 parts one at a time

Student 2

- Draws one whole and shades it, says he knows one whole is the same value as three-thirds
- Draws a second whole; begins the shade the entire whole
- Says, "Ooops", then explains he does not have another three thirds to make another whole. Redraws the second whole, partitions into three parts, shading two

Student 3

- Draws one whole and shades it, says he knows one whole is the same value as three-thirds
- Draws a second whole. Partitions second whole into two parts and shades both parts
- Explains he has to have five "parts" that fill wholes

FIGURE 5.4 Learner thinking for five iterations of one-third produce two wholes. Then, learner 3 could share that he initially thought that he had to "fill up" complete wholes with five parts as opposed to repeating ⅓ five times. Finally, learner 1 could share that he knew what the result of the iteration of ⅓ would be before drawing anything. Whatever you decide should depend on the mathematics goal you are looking to support and should work to position what is mathematical about each learners' solution. We suggest finding a way to incorporate all learners' approaches into the "Share" portion of the lesson.

Moves to Support Learners' Talk

Choosing a sequence ahead of time and planning out what kind of discourse may result support the discourse opportunities that learners are to have during "Share" time. Three moves (Chapin et al., 2009) often discussed in the mathematics education literature to support learner discourse are (a) re-voicing, (b) re-stating, and (c) applying reasoning. *Re-voicing* is a tool that teachers can use to promote mathematical discussion by saying the learner's exact words back to them. It is important not to change their words, as it might undermine the learner seeing their thinking as valid. Additionally, learners who have cognitive differences, such as working memory, benefit from hearing their own words mirrored back to them in order to reflect on their thoughts. *Re-stating* is a talk move used to elicit learner-to-learner conversation around a mathematical idea. It involves a request from the teacher for a learner to repeat or paraphrase what another learner has shared about his or her thinking. This is different from re-voicing because it engages learners in discourse with other learners as opposed to with the teacher. *Applying reasoning* promotes discourse around the critical thinking of others. The move occurs by asking questions such as "Do you agree or disagree?" and "Why?" (Hunt et al., 2018).

Let's revisit the learner thinking shown earlier. Consider a related task where learners were asked to share three churros among four friends. Suppose learner 3 draws the following representation and, when asked to share, taps the table twice (a long tap and a short tap) and mutters, "I shared them." To nurture more discourse, we need to extend the talk moves. Specifically, we can extend re-voicing to include *re-showing*. Re-showing means re-creating the pointing and tapping gestures that the learner used in his explanation; it is often used with re-voicing to provide a truer account of learners' reasoning. The benefit is that the learner is seeing and hearing their actions mirrored back to them so they have the opportunity to reflect on their thinking. Here is an example of a "re-show/re-voice":

> Let me see if I understand: You showed [pointing to each of the learner's drawn bars] and then you [as she taps twice] tapped the table two times. And that shared the three ice-cream bars for the four people. Is that what you said?

Re-show/re-voice positions learners' mathematical ideas as worthy. The notion of allowing learners to show (as opposed to only talk about) their thinking can also be combined with the structure itself (i.e., a think-pair-show-share to invite learners to make sense of this learner's solution with a partner).

Above, we also named "applying reasoning" as a possible talk move you can use to promote learner discourse. Suppose learners 1 and 2 are asked to compare their reasoning. Yet learner 2 does not want to share his thoughts in front of his classmates. To support the learner, you might consider asking learner 2 if he agrees or disagrees with learner 1's reasoning. If the learner does not respond, employ wait time by saying, "I'm going to pause and think – let's all do that." Afterward, learner 2 quietly states, "I don't know." At this point, you might agree and say that you, too, are unsure and ask all learners to partner up and compare both learners' reasoning. In the *pair-share-compare*, the two learners are able to compare each other's models more openly. In the paired discussion, the two learners discuss the similarities and differences in each of their representations. This alternate version of the pair-share parts of our intervention structure provides additional access by modeling both what it means to be unsure and how to persevere in problem-solving. It also provides all learners with opportunities to discuss their reasoning in small groups before being expected to participate in whole-group discussion.

Within each of these moves, we cannot stress enough the importance of wait time. Wait time means that you provide learners adequate time to respond to a question. Learners often require much more processing time than most teachers provide. Saying things such as "Let's wait for everyone to think this one through" or "I'm going to take a thinking pause on this one – let's all do that" can help all learners slow down and take the necessary time to engage with what is being asked.

Revisiting Social Norms

The opportunities that learners have to think about, pair and discuss, and share their reasoning in the larger group further build the norms of the *Small Environment*. As we explained in Chapter 4, "norms" in mathematics take two forms: (1) social norms and (2) sociomathematical norms (Cobb & Yackel, 1996). Remember that *social norms* are regularities in small group activity and are jointly constructed by both the learners and the teacher. Explaining and justifying solutions, making sense of explanations given by other learners, and agreeing and disagreeing on a particular solution are all examples of social norms.

Yet *sociomathematical norms* are linked to what learners count as being the same and different mathematically in the context of interacting in the small group. They include things like learners judging for themselves what counts as an acceptable mathematical solution, an insightful solution, or an efficient solution or linking different solutions to the same problem. Sociomathematical norms develop and are renegotiated over time in the context of the group space. They are considered to be the social component to learners' mathematical beliefs, values, and disposition. Like social norms, sociomathematical norms and learners' beliefs and positions influence each other and evolve over time as they are negotiated and renegotiated in the small group.

Working on Sociomathematical Norms over Time

Sociomathematical norms cannot shift until learners see their own reasoning as viable. That is why we have spent time in this book giving tools that you can use to position learners as competent mathematical thinkers and doers, such as developing social norms in the Small Environment. Once learners engage in social norms that position their own reasoning as viable, they can begin to construct sociomathematical norms during "Share" time in intervention lessons, such as noticing different pathways in viable mathematical solutions.

Teachers can help learners build sociomathematical norms. Supporting sociomathematical norms includes promoting learners to listen to each other's reasoning, direct comments, and normative statements. *Promoting active listening* involves direct questions posed by the teacher to engage learners in the reasoning of others and to judge the viability, similarities, and differences in the mathematics offered. (For example, "Who agrees? Who disagrees? Who thought about it in a different way? Convince me this way is different.") Generally, the teacher is prompting learners to engage or do something that promotes them to consider their own reasoning against that of other learners. *Direct comments* include publicly calling out the importance of norms and why they are used during "Share" time. *Normative statements* involve the teacher publicly praising learners who used sentence starters for using sociomathematical norms (e.g., marking when a learner engages in the reasoning of another). In the table below, we summarize strategies that teachers may use to support sociomathematical norms.

We conclude this chapter by inviting you to think through how each part of the intervention structure can further support and solidify the norms of your intervention group. See Table 5.5 for ideas. Being purposive and mindful of how norms adjust to each part of the structure will go a long way to ensure a successful implementation.

Putting It All Together

The final two chapters in this book – one in the classroom setting and one in the virtual setting – illustrate how the principles exemplified throughout this book can be put into action.

TABLE 5.5 Norms within Think, Pair, and Share

	Social Norms	*Sociomathematical Norms*
	Teachers might use these questions to drive co-constructed norms	*Learners might ask themselves these questions*
Think	• Do your learners need absolute quiet "Think" time? • Do they need to talk out loud to themselves in order to process the information? • Is there a requirement to show your thinking or is it just personal reflection? • How will you navigate the learners that want to just jump in and start talking? • What is the role of the teacher during "Think" time?	• How am I preparing my thinking to be shared with either a partner or the group? • What do I do if I get stuck? • Does my answer make sense?
Pair	• Who will determine who talks first? • What is the expected outcome of the conversation? • How will you ensure that learners are actively listening and learning from each other?	• How does my thinking compare to my partner? • What are the similarities in our thinking? • What are the differences in our thinking? • Does their answer make sense?
Share	• How do you sequence the learner thinking? • What is the expectation of the learners during this time? Will they be expected to write anything down?	• How does my thinking compare to my peers? • What are the similarities in our thinking? • What are the differences in our thinking?

Chapter 6 shares a Small Environment where learners were working together to bolster their conceptions of fractions as quantities (Hunt et al., in preparation). The chapter's aim is to give teachers a practical way to document learner thinking over time, making connections to the special education research and policy on tiered interventions. Chapter 7 reflects on the big ideas presented throughout this book and places them in the context of a newly added component to the educational landscape: the virtual environment.

References

Chapin, S., O'Connor, C., & Anderson, N. (2009). *Classroom discussions: Using math talk to help students learn, Grades K–6* (2nd ed.). Sausalito, CA: Math Solutions Publications.

Cobb, P., & Yackel, E. (1996). Constructivist, emergent, and sociocultural perspectives in the context of developmental research. *Educational psychologist*, 31(3–4), 175–190.

Hunt, J. H., MacDonald, B., Lambert, R., Sugita, T., & Silva, J. (2018). Think-pair-show-share to increase classroom discourse. *Teaching Children Mathematics*, 25(2), 78–84.

Hunt, J., & Silva, J. (2020). Emma's negotiation of number: Implicit intensive intervention. *Journal for Research in Mathematics Education*, 51(3), 334–360.

Hunt, J., & Tzur, R. (2017). Where is difference? Processes of mathematical remediation through a constructivist lens. *The Journal of Mathematical Behavior*, 48, 62–76.

Smith, M. S., & Stein, M. K. (2011). *5 practices for orchestrating productive mathematical discussions*. Reston, VA: National Council of Teachers of Mathematics.

6

How Do I Document and Use Learners' Changes in Thinking to Inform Instruction within Small Environments?

Small Environments in Action

In this chapter, we discuss how to document learners' thinking and use it to inform instruction. We will begin with a scenario of three learners within an intervention designed to bolster learners' conceptions of fractions. We will describe multiple tools that we used to determine learners' incoming understanding and how we designed intervention to leverage learners' strengths. Then, we describe a novel way to monitor learners' understanding throughout the intervention: learners' daily problem-solving strategies. Teachers can use learners' daily strategies alongside more common formative assessments to gauge changes in thinking over time and adjust instruction. Finally, we provide a few tips to further differentiate Small Environments in response to unique fraction understandings.

Bolstering Learners' Concepts of Fractions: A Case of One Small Environment

Meet Candice, Lili, and Nico, three fifth-grade learners whom we met with for a number of instructional sessions designed to bolster the learners' notions of fractions as quantities (Hunt & Martin, in preparation). We worked with our school partner to identify the three learners as those who would benefit from spending time in our Small Environment. Our work with the school began when we co-created what is often referred to as a "universal screener" for fractions to identify learners to work in the Small Environment.

Constructing and Using a Screener

Screeners are one tool that teachers can use to identify learners who may benefit from additional instructional time in Small Environments. Screeners are often multiple-choice exams that cover standards in a domain addressed during instruction throughout the academic year. Districts often create screeners, such as benchmark exams, for schools and teachers to administer to learners at varying points as a form of formative assessment. As a result, many schools and teachers have access to mathematics screeners that can be used in tandem with some of the tools mentioned in Chapter 3 in order to uncover learners' understanding of mathematics. Sometimes, districts do not have screeners and teachers create them.

In our case, the school did not yet have a screener. Yet they did have a standards-aligned exam, given at the end of each grade, that measured learners' yearly progress in mathematics. Because the test contained items that measured learners' fractions knowledge and was already validated, we created a screener from that test. In our case, we designed the screener to consist of fraction questions from the third, fourth, and fifth grades because we were interested in supporting learners in grade-level content as well as foundational concepts necessary to access grade-level content. More often, screeners are designed to measure only grade-level content. Teachers should consider both options and choose the one that better fits their instructional needs. (Will intervention be based on grade-level content only? Will it address learners' Individualized Education Program [IEP] goals? Will it take into account prior skills or concepts from the previous year(s)?) Below, we outline some of the information that we gained from the screener for the three learners in our Small Environment.

Candice, Lili, and Nico came to work with us because they all obtained a score below 50% on our screener. To use the screener data as a part of a multifaceted view of learners' mathematics (see Chapter 3), we looked at the items that each learner got correct and those that each learner was still working to understand. For example, all three learners correctly answered questions about fractions as representations of parts and wholes (e.g., "Given a rectangle partitioned into six parts and five parts shaded, name the fraction"). Additionally, all learners correctly answered questions that asked them the value of a point on a number line with an interval of zero to one, such as in Figure 6.1 below.

Whereas questions like these were answered correctly, others revealed concepts that learners were still thinking about. For example, two learners missed questions that addressed the meaning of unit fractions (e.g., identifying the fraction model that shows one third, such as a circle with six parts and two shaded, or identifying what fraction of a bar partitioned into six parts can be covered by two of those parts). Additionally, questions that addressed equivalent or multiplicative notions of fractions (e.g., the quantity that results from sharing one half of a pan of cake among four learners or the amount of pizza each of six friends obtains when equally sharing an eight-cut pizza pie) were missed by all three learners. As shown in Table 6.1, learners' answers to questions on the screener suggested a range of strengths as well as areas in which learners would benefit from additional instruction.

What is the distance from 0 to point B on the number line?

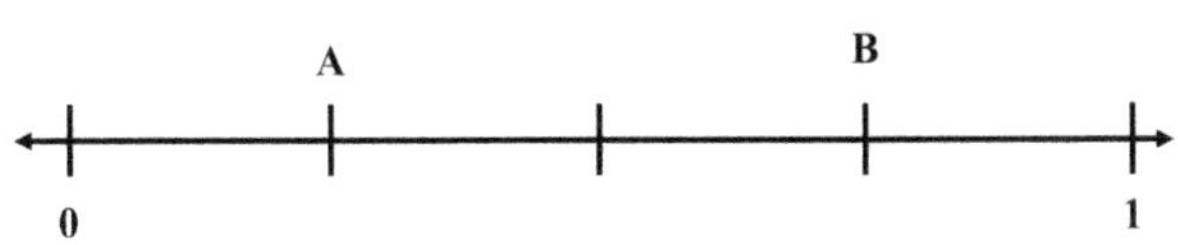

FIGURE 6.1 Example question on fraction screener

TABLE 6.1 Learner performance on fraction screener

Item	*Concept and standard addressed*	*Candice*	*Lili*	*Nico*
1	Part whole	✔	✔	✔
2	Part whole	✔	✔	✔
3	Equivalence	✖	✔	✔
4	Equivalence	✖	✔	✖
5	Unit fractions	✔	✔	✔
6	Unit fractions: Equivalence	✖	✔	✖
7	Unit fractions: Equivalence	✖	✔	✖
8	Comparing fractions: Equivalence	✖	✔	✖
9	Unit fractions: Non-unit fractions	✖	✔	✔
10	Unit fractions: Non-unit fractions	✖	✔	✔
11	Unit fractions: Non-unit fractions	✖	✔	✖
12	Part of a part: Multiplying fractions	✖	✔	✖
13	Multiplying fractions: Measurement model of fractions	✖	✔	✖
14	Part of a part: Multiplying fractions	✖	✖	✖
15	Part of a part: Multiplying fractions	✖	✖	✖
16	Multiplying fractions: Measurement division	✖	✔	✖
17	Measurement model of fractions	✖	✖	✖
18	Distributive model of fractions: Fractions as division	✖	✖	✖
19	Distributive model of fractions: Fractions as division	✖	✖	✖
20	Distributive model of fractions: Fractions as division	✖	✖	✖

Preparing for Small Environments – Uncovering Strengths and Prior Knowledge

Starting with screener data from the learners' performance in their mathematics classroom was helpful in determining their strengths and possible focal areas for interventions. Yet more information was needed: Although the screener gave us an idea of particular questions answered correctly or incorrectly, it told us little about the learners' thinking or why they answered the questions the way they did.

To dive deeper, we decided to gather three additional forms of information (see Chapter 3). First, we conducted informal interviews with each learner to better understand their prior knowledge. We were able to find out more about each learner's interests, activities they engaged in within their community, and what each learner identified as activities in which they excel. Second, we designed a short clinical interview to gain a richer understanding of each learner's notions of fractions. We designed the interview questions to address both the items from the screener (i.e., to invite learners to talk through their reasoning so we could better understand what they knew and areas they were still thinking about) and new, wider tasks that would allow a range of problem-solving strategies and representations. Finally, because the informal interviews helped us identify times during the school day when each learner felt successful, we were able to observe each of them twice; this gave us an additional understanding of each learner. Table 6.2 shows a summary of the information that we gathered for the learners. We wrote problems on paper and asked learners to contextualize each one. (For example, "Show the learner a paper with '2 share 5' written on and ask, 'What are we sharing today?'"). Take a look at the table and see what stands out to you.

Promoting Learners' Activity in the Small Environment

For us, three things stood out from which to consider a beginning for intervention. First, each learner brought in nuanced understandings of fractions. On the surface, this made it challenging to find a common starting place – a frequent and often stressful experience for many teachers. However, we see it as an opportunity to invite diverse ways of thinking and to allow these understandings to build upon each other. If our intervention had tasks that would invite each learner to actively employ their prior knowledge (Chapters 3 and 4), then they could compare and contrast their reasoning with each other (Chapter 5). Second, in the informal interviews, each learner reported that they enjoyed classes and subjects where they could talk to other

TABLE 6.2 Uncovering learners' thinking

	Candice	*Lili*	*Nico*
Informal Interview			
Competencies	<u>Favorite things to do in school</u>: Enjoys science the best because she gets to conduct science experiments and talk with her friends. Feels comfortable sharing with the teacher. <u>Favorite things to do after school</u>: Spend time with her mom and family. Family owns a small chain of bakeries in the community.	<u>Favorite things to do in school</u>: Says that she enjoys most of her classes. She really likes classes where she can talk with her friends and the teachers about the content. <u>Favorite things to do after school</u>: After-school club and spend time with her family. She belongs to several after-school clubs and talks a lot about a debate that is an activity in one of the clubs.	<u>Favorite things to do in school</u>: History is his favorite topic because he likes to hear about different facts and dates. He also enjoys creating projects while working in groups. <u>Favorite things to do after school</u>: Play soccer and spend time with his teammates.
Disposition for Math	Describes math as "hard" and says that she dislikes the subject. She does say that she does better in math with help from the teacher.	Describes math as "OK." When I asked for more explanation, she shared that she doesn't get to work with her friends as much or share her thinking, so she finds math a bit boring.	Describes math as "boring" and hard. Says math is not something that he is good at; when I asked for more, he said that he "just doesn't think of math as something he likes."
Home/ Community	Named various restaurants around the school she has visited with her family. She talks about different foods she ordered and named some favorites. Says that her family uses math at their bakery.	Talks about playing games with her siblings at home and helping cook dinner with her dad. She also talked about walking a trail with her family on the weekends. She says that she can think about math as walking a number of miles with her family.	Talks a lot about the soccer fields near the school and in the community, how he plays soccer there, and at times shares snacks (granola bars; Gatorade) with his teammates. Can't think of anywhere where he actively uses math outside of school.

Clinical Interview

Screener Problems	1. What fraction is shown in the figure?  a. ⅓ b. 4/6 c. 2/4 A Explained 4/6 was the answer "because two are shaded and four are not shaded and there are six". Then said that the answer was 2/6. When I asked if 2/6 could be expressed in another way, she said no. When I asked how many times the shaded blocks would need to be used to make up the total of six parts, she said "two." 2. Share one half of a pan of cake among four learners. How much of the whole cake does each learner get? a. ¼ b. ⅛ c. ½	1. What fraction is shown in the figure? 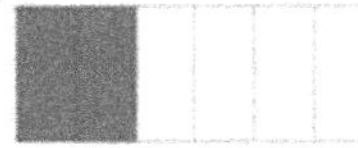 a. ⅓ b. 4/6 c. 2/4 A [immediately] "One third because it takes three 2 s to make six." 2. Share one half of a pan of cake among four learners. How much of the whole cake does each learner get? a. ¼ b. ⅛ c. ½ A "Well, if they each get the same then it's a fourth of the half." T "How much of the whole cake is that?" A "Well, I think 2/4 because there are two halves in a whole and they have one fourth."	1. What fraction is shown in the figure?  a. ⅓ b. 4/6 c. 2/4 A Explained 2/4 was the answer because two are shaded and four are not shaded. 2. Share one half of a pan of cake among four learners. How much of the whole cake does each learner get? a. ¼ b. ⅛ c. ½ A "It's a piece of the half. One piece for each." T "What would we call the size of that piece?" A "One?"

(Continued)

TABLE 6.2 (Continued)

	Candice	*Lili*	*Nico*
Informal Interview			
	A "Um, I'm not sure. I think maybe one fourth because there are four people." T "OK. How much of the whole cake is that?" A "I really don't know."		
"Wide" tasks	2 share 5 • Conceptualized as cookies and friends. Gave two cookies to each friend. Cut the final cookie into somewhat equal parts. Called the amount for each person "three." 4 share 3 • Cut all of the cookies into halves. Gave out one half for each. Then cut the final bar into halves again. Said each person got "One half and a piece." 2 and/or 3 share 1 • Folder the bar into halves. Called each part one half. • Said there was no way to create three equal parts in the bar.	2 share 5 • Immediately said "two and one half. I just know it." 4 share 3 • Said to cut all of the items into four parts, because 4 x 3 = 12. Said each person got 3 parts, which is 3/12. With a prompt, called the amount ¾. 2 and/or 3 share 1 • Folder the bar into halves. Called each part one half. • Folded the bar and said there would be four parts now before unfolding it because 2 x 2 = 4. Then used a pencil to estimate the size of one bar, then repeated it to make three bars. Called the part one third.	2 share 5 • Conceptualized as granola bars and teammates. Said each would get two whole bars and one part of the final bar. Called the amount for each person "two whole bars and one out of two parts of the last bar." 4 share 3 • Cut all of the cookies into halves. Gave out one half for each. Then cut the final bar into halves again. Said each person got "One half and a fourth." 2 and/or 3 share 1 • Folder the bar into halves. Called each part one half. • Tried to fold the bar into three parts. Then said there was no way to create three equal parts in the bar.

	1/2 x 5 = • Drew five bars, then cut each bar into two parts. Called the quantity "Ten." 2/3 x □ = 6 • Child could not solve the problem.	1/2 x 8 = • Said the quantity would be "eight halves, which is the same thing as four." 2/3 x □ = 6 • Drew six bars, cut each bar into three parts. Began to color in two parts in each bar. Said, "Six?" Paused, then colored in one part from each bar until all parts were colored in. Said, "Six… seven [pointed to two bars], eight [pointed to two bars], nine [pointed to two bars]. It's nine.	1/2 x 5 = • Drew five bars, then cut each bar into two parts. Colored in the first two bars, then one part of the final bar. Called the quantity, "two whole bars and one part of the last bar." 2/3 x □ = 6 • Child could not solve the problem.
Observation			
Participation	Candice participates verbally and with gestures during science class. She explained the results of an experiment for her team and engaged in conversation with other peers about the work.	Lili took part in a mock debate after school. Lili willingly shared the visual representations that she prepared in planning for the debate and referenced them as she spoke.	Nico showed leadership during the soccer match. He led a conversation with his teammates about a planned play/strategy and worked with his team to execute it on the field.
Showing Competence	Observation: Science class Candice participates verbally and with gestures during science class. She often raises her hand to contribute when asked by the teacher and shares frequently as learners present projects.	Observation: After-school club Lili participates verbally and visually. Lili prepared wonderful visual representations which she shared with her group and often referenced as she and her peers prepared their half of the debate.	Observation: Recess Nico participates verbally and physically with his peers while playing volleyball. He talks frequently with his peers as he plays ("I'll set!"; "Mike, hit that over!").

learners about their thinking yet also appreciated the support of the teacher at times when thinking through a problem. This was also evident in observations of each learner (e.g., Candice in science class explaining the results of an experiment, Lili in an afterschool activity where learners engaged in a mock debate, and Nico during soccer where he worked collaboratively with his teammates to execute a planned play/strategy). Finally, learners did not seem to view mathematics in the same light as they did other subjects or connect some of their strengths to math time (i.e., planning a strategy or explaining or justifying ideas).

From the information that we gathered, we were able to begin to design tasks to use in our Small Environment. We began by designing several days of Accessing tasks (see Chapter 4). Recall from Chapter 4 that Accessing tasks are wide tasks that have many entry points, which allow learners to find many different solutions. They also help the teacher further deduce the learners' thinking. As a reminder, it is a good idea to begin with several "Accessing" tasks if teachers find themselves beginning instruction without a good handle on the prior knowledge of their learners. Beginning a task sequence with at least three sessions' worth of Accessing tasks is standard in terms of task selection and sequencing. As explained in Chapter 5, we employed a Think-Pair-Share structure to support learners' active problem-solving, reflection, and discussion of their reasoning for each of the first four days of instruction. Additionally, we spent the entire first day and portions of the time in the other three days to construct group norms with learners to guide math time. In Table 6.3, we list the tasks that we used each day, the prompts and discourse moves that we employed (generally and for each learner), and examples of learners' work for each day of instruction.

Assessing Learners' Initial Thinking – Instructional Decision 1

As we mentioned above, our first four days of instruction consisted entirely of co-constructing norms and engaging in Accessing tasks, which for us were equal-sharing tasks (i.e., sharing multiple items among multiple people). We also decided to use our original equal-sharing tasks as a formative gauge of learners' evolving understandings, a gauge that we would

TABLE 6.3 Days 1–4 in the Small Environment

Prompts used throughout	Re-voice: "You said…. Is that what you said?" Cause and effect: "What happened when you?... Is that what you thought would happen?"		
Talk moves used throughout	Re-voice: "You said…. Is that what you said?" Re-state: [Ask learners to put other learners' reasoning in their own words.] Apply reasoning: "Do you agree/disagree? Why"		
Day 1 and throughout	Co-construction of norms		
	Day 2 4 share 9	Day 3 3 share 5	Day 4 6 share 4
	Four learners shared nine ice cream bars equally. They finished all of the ice cream bars. How many ice cream bars did each learner eat?	*Three soccer players want to share five large pizzas so that each player gets the same amount to eat. How much pizza does each player get to eat?*	*Six friends want to share four sticks of clay for an art project. How much clay can each friend have if they all get the same amount?*
Candice		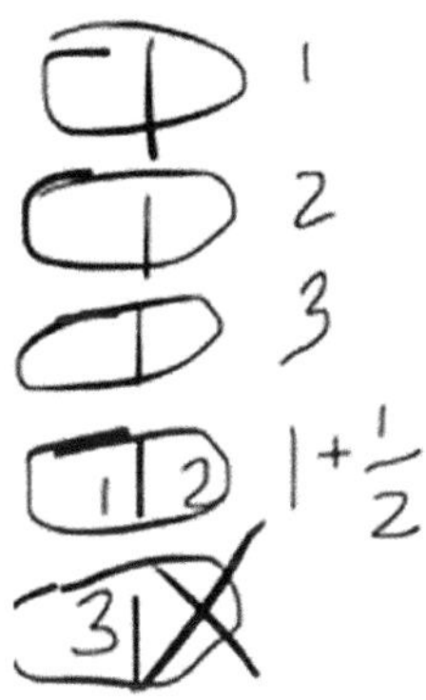	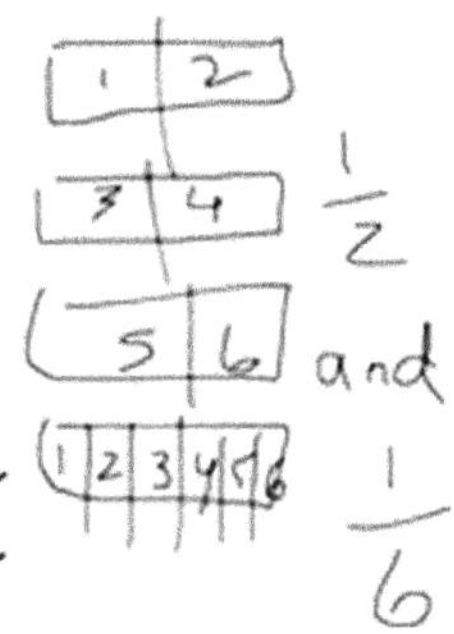
Nico	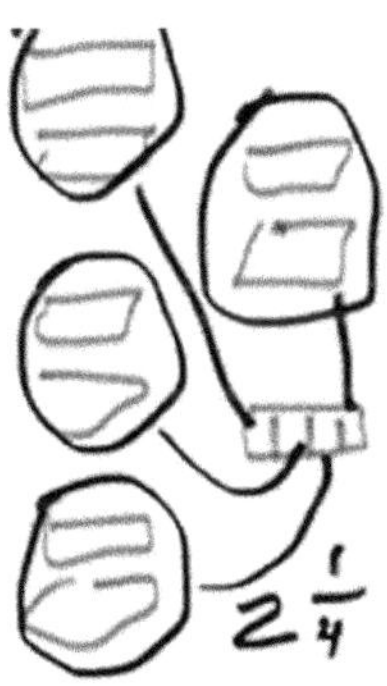	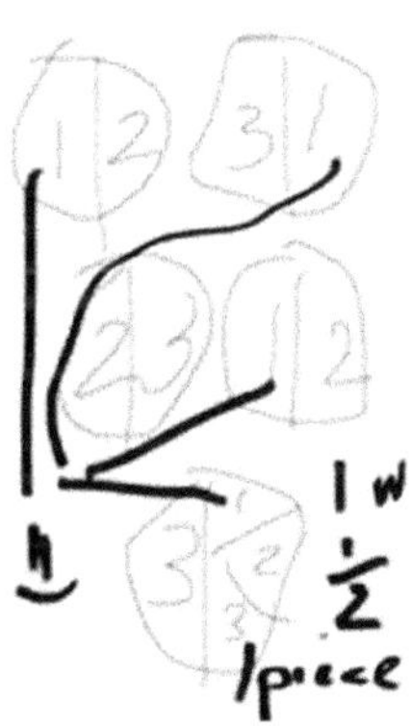	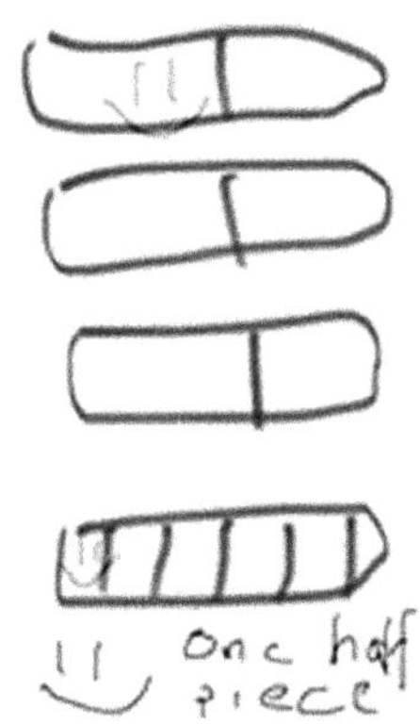

(*Continued*)

TABLE 6.3 (Continued)

Lili	"It's two wholes and one fourth of the last whole."	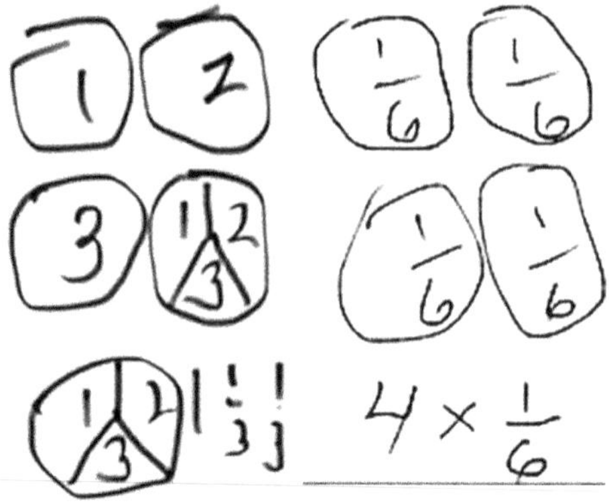

continue to employ beyond the initial sessions. We chose to do this because learners' strategies for solving problems are a window into their thinking; they provide teachers an observable and documentable way to understand learner thinking over time (e.g., Hunt et al., 2016; Siegler, 2007). Documenting learners' strategies in the equal-sharing tasks over the first four sessions gave us a clearer understanding of each learner's thinking (see Table 6.2). Moreover, we had additional understandings of what kinds of tasks might work to bridge learner understanding toward our overall learning goal, which was to bolster learners' understanding of fractions as quantities.

To better understand what that means, we recommend that teachers consult resources that unpack trajectories, or progressions, of learner understanding. To some extent, state and national standards reflect such a progression. For example, revisiting the performance information from the learners' screeners, we believed that two learners were still thinking about the idea of *unit* fractions as quantities while another learner already had that understanding. The learners' work in the first four days of intervention confirmed our beliefs. Namely, Candice and Nico thought about fractions using a "Halves" conception whereas Lili thought about fractions through an "Anticipatory Partitioning" conception. Learners who think with a "Halves" conception are building up an understanding of unit fractions as related to a whole. Conversely, learners who think with an "Anticipatory Partitioning" conception have a basic understanding of unit fractions inside the boundaries of a whole (e.g., an understanding of ⅗ as three iterations of ⅕) and are working to understand other unit fractions beyond the boundaries of one whole, such as 7/5. The two ways of reasoning are different yet can inform each other through instruction (see Chapter 5). The graph below shows the learner thinking in relation to one developmental progression that we often consult to gauge learners' fraction reasoning (Hunt et al., 2020). The developmental progression appears below the graph.

Bridging and Promoting the Need for Change

Using the learner data, we made an instructional decision to add two tasks (Hunt et al., 2020) to our Small Environment: using a unit fraction to create non-unit fractions (Empson & Levi, 2011) and creating and adjusting the magnitude, or length, of a unit fraction in reference to the length of a whole (Tzur & Hunt, 2015). The first task (see Table 6.3, day 5) is useful to help learners see that unit fractions can be used to make other fractions – both within and beyond the bounds of one whole (Empson & Levi, 2011). It also helps learners to re-form wholes and distinguish wholes from unit fractions.

In this way, the task becomes both a bridge and a challenge for Lili (i.e., she already distinguishes parts and wholes yet works to understand improper fractions as an iteration of unit fractions and wholes and as a quantity itself) and for Candice and Nico (i.e., they have working knowledge of the relation between halves and wholes yet are working to build that same relation between other unit fractions and wholes along with unit fractions and related small non-unit fractions inside a whole). We planned to keep created non-unit fractions under one (Candice and Nico) or over one (Lili).

The second task (day 6) is a bridge for learners who think like Candice and Nico. They see the inverse relationship of unit fractions to wholes when creating halves, but thinking that relies on halving becomes problematic when creating other unit fractions, such as thirds or fourths. Although the task is likely not a bridge or a challenge to learners who think similarly to Lili, it could help solidify their understanding of the relation as multiplicative (i.e., 5/5 as five times as large as ⅕) with less common unit fractions. Figure 6.2 shows learners' progression with the tasks across the

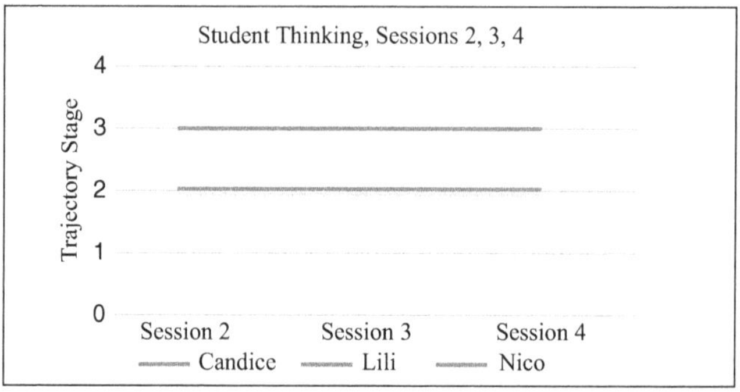

Trajectory Stage	Divisibility of the Whole	Partitioning Plan	Relation of Unit to While and Iteration
(0) No Fractions (thinking about counting/WN)	**No Fractions** Will only share/deal out wholes. Whole not yet conceived as divisible. Does not act on the whole or create fractions		
(1) Emergent Sharer (Comes in with 1 level of Units)	**Developing** Seems to cut item or items into pieces reluctantly	**Developing** Trial and error based in whole number in activity. • Partitioning across whole and/or leftovers is difficult • May begin to use "half" in activity, but it is not meaningful to them as a quantity	**Developing** Student attends to making parts or using the whole. • Parts created are not equal in size and the student is not bothered. • Parts are equal in size, yet whole is not exhausted.
(2) Using Half (2 levels of Units in action)	**Solidified** Readily divides whole without hesitation.	**Developing** Plan becomes evident in dealing with the leftover in activity. • "Half" represents a meaningful quantity used of partition. • May link number of pieces to number of sharers	**Developing** Begins to coordinate equal parts in the whole "after the fact" when dealing with leftover. • Pays close attention to creating equal size parts AND using all of the item or items.
(3) Anticipatory Partitioning (Comes in with 2 levels of Units)	**Solidified (*within one whole*)** Plans to create number of parts equal to number of sharers prior to activity. • Parts are planned within the wholes. • May use knowledge of multiplication or division to plan a number of parts.		**Solidified (*at two levels*)** Creates equal parts while exhausting the whole. • Justifies the value of a created part as the same as all other parts needed to recreate one whole. • Uses a created part to tests its size against a referent whole AND remake the size of the whole.
(4) Composite & Iterative Fractions (2 levels of Units/3 levels of Units in action)	**Developing (*within and across wholes & at three levels*)** • Uses a unit fraction (part) to count within (forming non-unit fraction) and outside of wholes Reform wholes through repeating/counting unit fractions. • Quantifies in terms of one whole using addition and subtraction (like denominator fractions) and multiplication (whole number multiplied by fraction). • Considers equivalent situations (e.g. nine-fifths is the same as one whole and four one -fifths.)		
Reversible Fractions (2 levels of Units/ 3 levels in action)	**Solidified (*within one whole*)** • Uses reversible notoin of non-unit fractions within one whole (e.g. 2/3 as 1/3 + 1/3) to reform wholes. o Undo iteration of m/n by partitioning m/n to create 1/n o Uses 1/n to undo partition of whole by iterating 1/n n time to make n/n		

FIGURE 6.2 Candice, Lili, and Nico across three instructional sessions

first four sessions and the learning trajectory that we were using to guide our instructional decisions. Study the graphs and consider what you notice.

Promoting Connections – Instructional Decision 2

The decision to include two additional tasks in our Small Environment proved advantageous in terms of both the changes in learner reasoning and the connections within learner reasoning. Table 6.4 shows the tasks that we used and the learners' work for each day. It is important to note that we rotated between problems that used a unit fraction to create non-unit fractions and the original equal-sharing problems each day of instruction. All days of instruction used the task where learners created and adjusted the size of a unit fraction as a short "game" for 5 to 10 minutes at the end of each session (for the task and game, see Tzur & Hunt, 2015). In these sessions, we used a choice of numbers to differentiate instruction for tasks. For example, on day 5, when learners were using unit fractions to make other fractions, learners could solve the problem with one fourth of a sandwich and *either* three or 12 friends. At times, we make the choice for learners if we are certain of which choice would align the task with our instructional goal (e.g., Bridging, Challenging, and Varying). In this instance, we chose three friends for Candice and Nico and 12 friends for Lili to facilitate a bridge for each learner.

As shown by the data, on day 8, all three learners were using at least Anticipatory Partitioning in the equal-sharing problem; Lili often named the equal share in terms of one whole without any partitioning and was beginning to make connections between the equal share in terms of one whole and the equal share in terms of all of the wholes. Candice and Nico were consistently using the number of sharers as an input into how they cut up each item. All three learners successfully repeated a unit fraction to make another fraction and were able to make connections between quantities that extended beyond a whole as both fractions greater than one and mixed numbers. Additionally, all three learners easily created and adjusted the magnitude of unit fractions within the length of a whole. Figure 6.3 shows the learners' progression with the tasks across the four sessions and

TABLE 6.4 Days 5–8 in the Small Environment

Prompts	Re-voice: "You said…. Is that what you said?" Cause and effect: "What happened when you?... Is that what you thought would happen?"			
Talk moves	Re-voice: "You said…. Is that what you said?" Re-state: [Ask learners to put other learners' reasoning in their own words.] Apply reasoning: "Do you agree/disagree? Why"			
Tasks	Day 5 Each friend eats one fourth of a sandwich for lunch. There are [3, 12] friends. How many sandwiches did the friends eat together? ----------- 2 learners share one French fry equally	Day 6 7 children equally share 4 bananas for a snack. How much of one banana does each child get? --------- 3 and 4 learners share one French fry equally	Day 7 Each friend eats one eighth of a sandwich for lunch. There are 13 friends. How many granola bars did the friends eat together? --------- [5, 8] learners share one French fry equally	Day 8 9 aliens share 3 food bars on a mission. How much of one food bar does each alien eat? How much of all of the food bars does each alien eat? -------- [6, 10] learners share one French fry equally
Candice	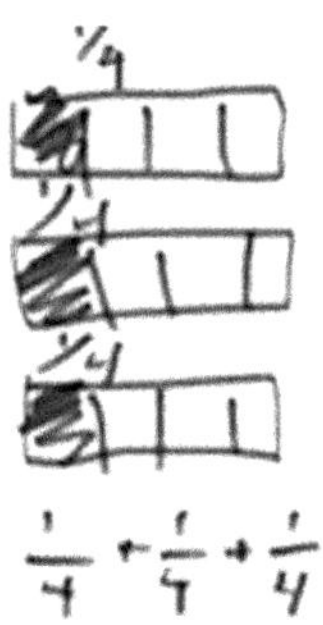 ----------- *Folded the fry into two parts and directly compared the lengths*	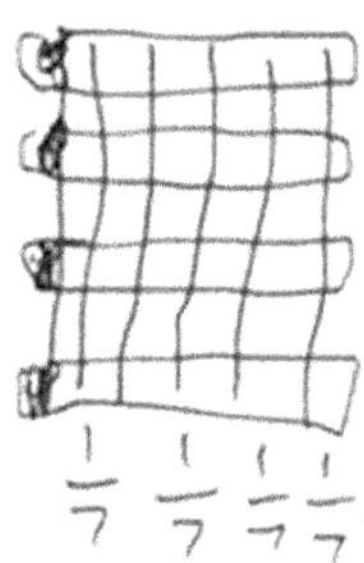 ----------- *Used repeat strategy to create and test a share length. Achieved correct length for three sharers after 7 attempts.*	 ----------- *Incorrectly made share longer, then made shorter. Achieved correct length for four sharers after 4 attempts.*	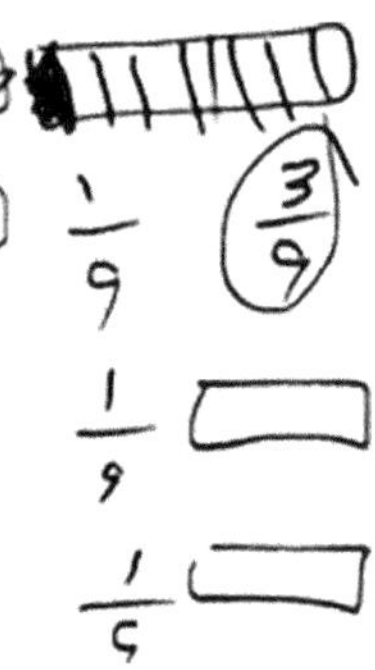 ----------- *Correctly adjusted length (shorter) for 5sharers. Correct adjusted relative amount. Achieved correct length for four sharers after 3 attempts*

(*Continued*)

TABLE 6.4 (Continued)

Nico	 ----------- *Folded the fry into two parts, pointed to one part and said, "each gets one half of the fry"*	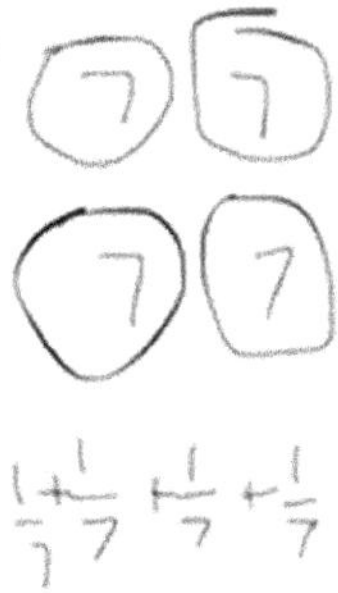 ----------- *Used repeat strategy to create and test a share length. Achieved correct length for four sharers after 2 attempts.*	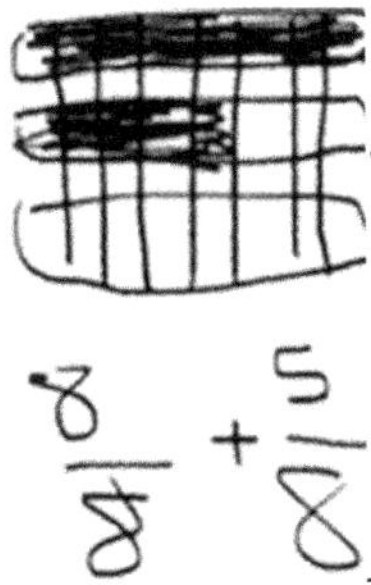 ----------- *Correctly adjusted length (shorter) for 4 sharers. Achieved correct length for four sharers after 4 attempts.*	 ----------- *Correctly adjusted length (shorter) for 8 sharers. Correct adjusted relative amount. Achieved correct length after 4 attempts.*
Lili	"Four, eight, twelve. So that's three wholes." ----------- *Said, "Each would get one half of the fry" and folded into two parts to show the share size.*	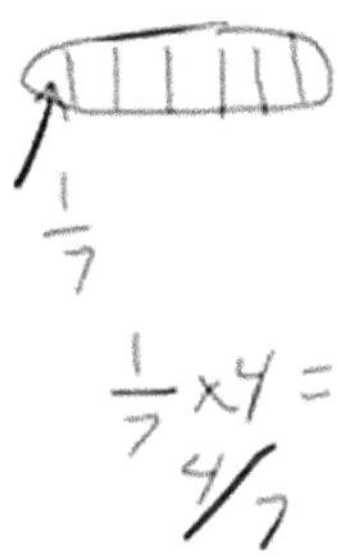 "28/7 across all 4." ----------- *Used repeat strategy to create and test a share length. Achieved correct length for four sharers after 2 attempts.*	"Thirteen one eighths or, 13/8" ----------- *Correctly adjusted length (shorter) for 8 sharers. Achieved correct length for four sharers after 2 attempts.*	 ----------- *Correctly adjusted length (shorter) for 10 sharers. Correct adjusted relative amount. Achieved correct length after 1 attempt.*

the learning trajectory that we were using to guide our instructional decisions. Study the graphs and consider what you notice.

We noticed that Candice's strategies in the equal-sharing problem changed over time. The change occurred as she played a game where she created and adjusted the size of unit fractions

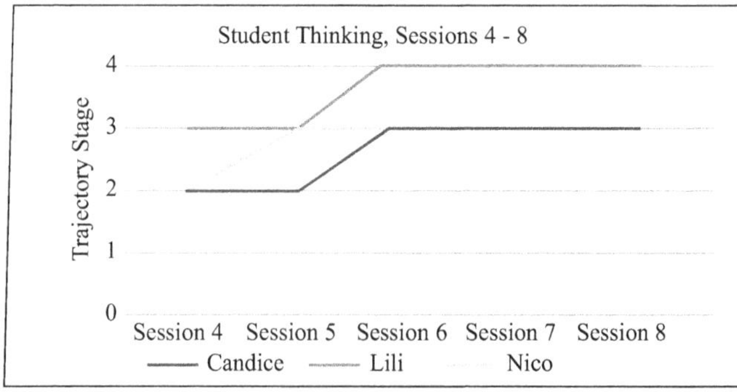

Trajectory Stage	Divisibility of the Whole	Partitioning Plan	Relation of Unit to While and Iteration
(0) No Fractions (thinking about counting/WN)	**No Fractions** Will only share/deal out wholes. Whole not yet conceived as divisible. Does not act on the whole or create fractions		
(1) Emergent Sharer (Comes in with 1 level of Units)	**Developing** Seems to cut item or items into pieces reluctantly	**Developing** Trial and error based in whole number in activity. • Partitioning across whole and/or leftovers is difficult • May begin to use "half" in activity, but it is not meaningful to them as a quantity	**Developing** Student attends to making parts or using the whole. Either: • Parts created are not equal in size and the student is not bothered. • Parts are equal in size, yet whole is not exhausted.
(2) Using Half (2 levels of Units in action)	**Solidified** Readily divides whole without hesitation.	**Developing** Plan becomes evident in dealing with tthe leftover in activity. • "Half" represents a meaningful quantity used of partition. • May link number of pieces to number of sharers	**Developing** Begins to coordinate equal parts in the whole "after the fact" when dealing with leftover. • Pays close attention to creating equal size parts AND using all of the item or items.
(3) Anticipatory Partitioning (Comes in with 2 levels of Units)	**Solidified (*within one whole*)** Plans to create number of parts equal to number of sharers prior to activity. • Parts are planned within the wholes. • May use knowledge of multiplication or division to plan a number of parts.		**Solidified (*at two levels*)** Creates equal parts while exhausting the whole. • Justifies the value of a created part as the same as all other parts needed to recreate one whole. • Uses a created part to tests its size against a referent whole AND remake the size of the whole.
(4) Composite & Iterative Fractions (2 levels of Units/ 3 levels of Units in action)	**Developing (*within and across wholes & at three levels*)** • Uses a unit fraction (part) to count within (forming non-unit fraction) and outside of wholes • Reform wholes through repeating/counting unit fractions. • Quantifies in terms of one whole using addition and subtraction (like denominator fractions) and multiplication (whole number multiplied by fraction). • Considers equivalent situations (e.g. nine-fifths is the same as one whole and four one -fifths.)		
Reversible Fractions *(2 levels of Units/ 3 levels in action)*	**Solidified (*within one whole*)** • Uses reversible notoin of non-unit fractions within one whole (e.g. 2/3 as 1/3 + 1/3) to refor wholes. o Undo iteration of m/n by partitioning m/n to create 1/n o Uses 1/n to undo partition of whole by iterating 1/n n time to make n/n		

FIGURE 6.3 Candice, Lili, and Nico across five additional instructional sessions

within a whole. Nico's data suggests similar trends. Lili's data suggest that her strategies in the game were rather sophisticated and consistent from the beginning. Her thinking when using a unit fraction to create non-unit fractions outside of one whole also seemed sophisticated. Yet, on the 10th day, we challenged Lili to consider the number of times a recipe that called for ⅔ of a cup of flour could be made from 3 cups of flour. After that, she solved an equal-sharing problem on the 11th day and named the share in terms of all of the sharers. Lili may have made a connection between the number of times a fraction repeats to make a number of wholes and the result of equal-sharing division. In summary, the thinking of all three learners grew in sophistication yet remained nuanced. Particularly, Lili's thinking (advancing from Anticipatory Partitioning to a form of Distributive Reasoning) remained separate from that of Candice and Nico (advancing from "Halves" to "Anticipatory Partitioning" to a form of Iterative Reasoning).

Unique Thinking – Instructional Decision 3

Because the equal-sharing tasks and creating non-unit fractions tasks proved to support changes in thinking for all three learners, we continued with their use in the next four instructional sessions. However, we discontinued the use of creating and testing the magnitude of a unit fraction task. The task seemed to support Nico and Candice in employing more advanced partitioning strategies; all three learners in the task reflected more sophisticated levels of thinking after four days. We planned sessions 9 through 12 to attend to the nuances in each learner's thinking. Specifically, we intentionally planned differentiation into the tasks. We did this in three ways.

First, we gave choices in the number values for the creating non-unit fraction tasks to allow learners to either continue working with unit fractions within and outside of wholes or extend the work by using non-unit fractions to create other larger non-unit fractions coordinated with wholes. Second, we included questions in the equal-sharing problems that further supported connections to distributive reasoning (i.e., "How much of one

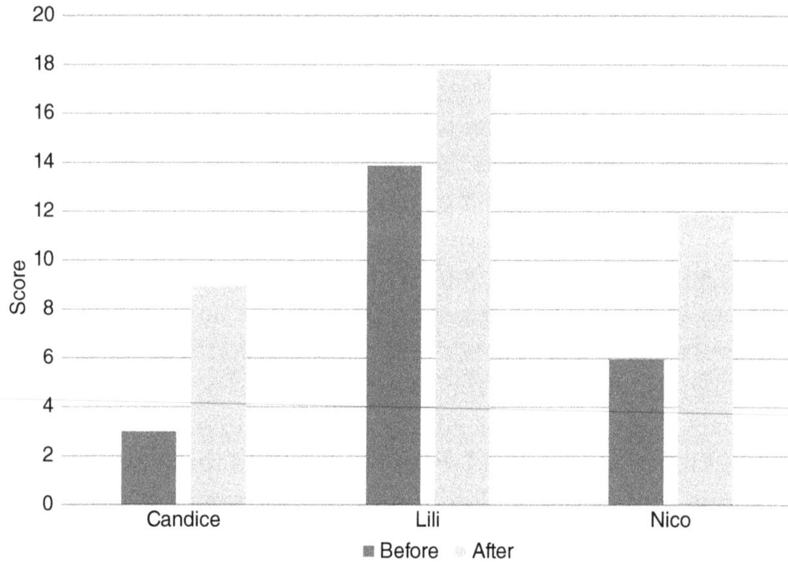

FIGURE 6.4 Performance changes before and after the intervention

whole?" and "How much of all of the wholes?"). Finally, we included tasks that invited other forms of multiplicative reasoning and equivalence (e.g., "Four people share 1 ½ bars of clay for a project. How much of one bar of clay for each person?").

As a more standard means of formative assessment, we re-administered our screener after session 12 to see whether performance improved over the first 12 sessions of intervention. As shown in Figure 6.4, we saw changes in learners' responses to questions that measured multiplicative notions of fractions that earlier were missing (e.g., the quantity that results from sharing one half of a pan of cake among four learners and the amount of pizza each of six friends obtains when equally sharing an eight-cut pizza pie). Together, the daily problem-solving strategies and changes in learner performance suggested that our Small Environment was effective thus far in supporting advances in learners' understanding of fractions as quantities.

Conclusion

In this chapter, we supplied some ideas to document learner thinking and use it to inform instruction. We described multiple tools that teachers might consider to determine learners' incoming understanding. Namely, we described using learners' daily problem-solving strategies alongside more common formative assessments to gauge changes in thinking over time and adjust instruction. Understanding learner thinking in this way can help teachers adapt tasks and other aspects of instruction to support learners to extend their reasoning.

References

Empson, S. B., & Levi, L. (2011). *Extending Children's Mathematics: Fractions and decimals*. New York, NY: Heinemann.

Hunt, J. H., Martin, K., Khounmeuang, A., Silva, J., Patterson, B., & Welch-Ptak, J. (2020). Design, development, and initial testing of asset-based intervention grounded in trajectories of student fraction learning. *Learning Disability Quarterly*, 0731948720963589.

Hunt, J. H., Westenskow, A., Silva, J., & Welch-Ptak, J. (2016). Levels of participatory conception of fractional quantity along a purposefully sequenced series of equal sharing tasks: Stu's trajectory. *The Journal of Mathematical Behavior*, 41, 45–67.

Siegler, R. S. (2007). Cognitive variability. *Developmental science*, 10(1), 104–109.

Tzur, R., & Hunt, J. (2015). Iteration: Unit fraction knowledge and the French fry tasks. *Teaching Children Mathematics*, 22(3), 148–157.

7

How Do the Principles of the Book Apply to Virtual Small Environments?

Learner-Driven Intervention Instruction in a Virtual Space

When this book was being written, the world, and more specifically the landscape of the education world, changed when the coronavirus pandemic hit. Prior to the pandemic, most teaching and learning occurred in person. However, owing to the need to reach learners virtually, we now think differently and have adapted to the "new normal." The adaptation has afforded opportunities for us to consider what it means to create "Small Environments" in an equitable, virtual space. When thinking back over the information presented in this book, you may wonder: which of these principles and frameworks work well if instruction is not face-to-face? We would argue that all of them can be modified or adapted to the virtual environment. Our aim in this chapter is to support you to do that.

Historical Context in a Virtual Space

In Chapter 1, we focused on shifting from teacher-driven deficit-focused interventions to a learner-driven asset-based intervention model. A review of the historical context that supports traditional views of intervention allowed us to show that performance gaps are not actually gaps in learner performance but a gap in the *opportunities* that learners are given to grow what they know. Learners need opportunities to think and reason through mathematics from a place that makes sense to them in order to advance their thinking.

In times of pandemic, the populations that are most vulnerable have the most to overcome. In our experience during this time, some of the barriers are a lack of reliable internet or devices (or both) as well as concerns around attendance, most notably because students are having to play a more active role in supporting their family structure during traditional school hours. Therefore, their opportunities to learn may become a bigger barrier than in non-pandemic times. With these barriers in mind, we need to try our best whenever we have an opportunity to be with our learners and make every possible attempt to ensure that each of them has instructional time to think and reason around mathematics.

What It Means to Know and Learn Mathematics in a Virtual Space

In Chapter 2, we looked at how to use the five strands of mathematical proficiency in order to illustrate the complexity of a learner's mathematical understanding as well as their attitudes toward engaging in mathematics (see Figure 7.1). In conversations with teachers around virtual learning, when time is in shorter supply than before, they feel the pressure to focus more on the procedural fluency of mathematics because it is easier to recognize in a virtual environment. However, we want to impress upon you to look at these strands for their beautiful interconnectedness instead of thinking that you have to pick just one.

A simplistic example might be a lesson on dividing fractions. It might be easy to teach learners that "all you need to do" is "keep change flip." You might even think that you would

The Five Strands of Mathematical Proficiency	
Conceptual Understanding	comprehension of mathematical concepts, operations, and relations
Procedural Fluency	skill in carrying out procedures flexibly, accurately, efficiently, and appropriately
Strategic Competence	ability to formulate, represent, and solve mathematical problems
Adaptive Reasoning	capacity for logical thought, reflection, explanation, and justification
Productive Disposition	habitual inclination to see mathematics as sensible, useful, and worthwhile, coupled with a belief in diligence and one's own efficacy.

FIGURE 7.1 NRC (2001) five interconnected strands of mathematics of proficiency

increase their engagement by introducing the lesson through a catchy song or video. However, learners need time to grapple with the complexities of understanding that division, just like whole number division, has both partitive and quotative division structures (Carpenter et al., 2017). Learners need experiences in engaging in the differences between the two structures and connecting them to scenarios that they might encounter in their everyday life. We would encourage you to do this through a series of wide tasks, as shown in Figure 7.2. We will review wide tasks a little later in this chapter.

Chapter 2 also laid out how learners best learn mathematics when it is built using ideas they already have, is presented in a way that requires them to actively think through a problem, and prompts them to reflect on their thinking and the thinking of others. These *three learning mechanisms* were connected to the *three key teaching moves* over the next three chapters, as shown in Table 7.1. Whether you are teaching in person or virtually, these learning mechanisms and connected teaching moves do not vary. How some teaching moves are implemented may look different but their principles remain unchanged. In the next few paragraphs, we will unpack how you may use the teaching moves in a virtual space.

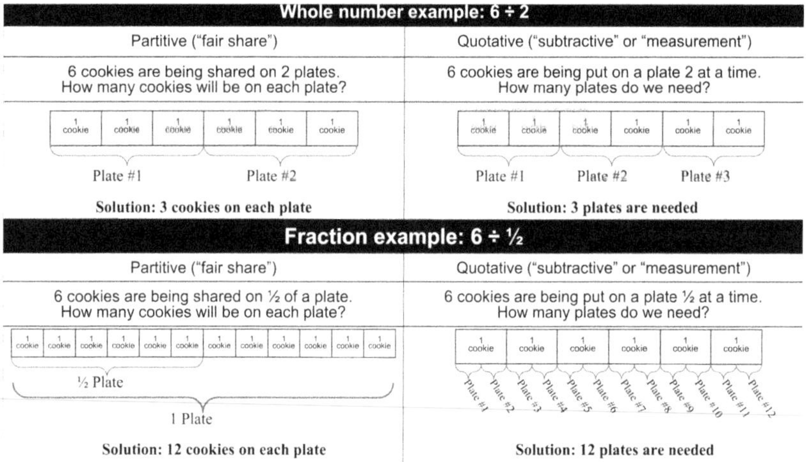

FIGURE 7.2 Partitive and quotative division structures

TABLE 7.1 Connecting teaching to learning

Learning mechanisms (learner actions)	*Key teaching move (teacher actions)*
Using ideas one already has	#1: Bringing forward learners' prior knowledge
Actively thinking through a problem	#2: Promoting *learners'* participation and thinking and the need for change
Reflecting on thinking and how it connects to others' thinking	#3: Promoting noticing, reflecting, and connections through purposive teaching

Bringing Forward Learners' Prior Knowledge in a Virtual Space

In Chapter 3, we looked at ways to learn more about your learners in order to bring forward their prior knowledge. We recommended using observation, conversation, and task-based interviews in order to uncover both the broad contextual knowledge and specific mathematical knowledge that a learner brings with them. In a virtual environment, conducting observations may be a barrier. However, some teachers have shared with us that they have gained a unique perspective into their learners' daily lives while being in a virtual environment. A learner's family members, including pets, may walk by or even "pop" into the class. Although this may seem disruptive, you may choose to view it as a literal window into your learners' lives. Some teachers have shared their surprise that learners who were not so communicative in person were more open in the virtual environment. One advantage of the virtual environment is that it often removes the social pressures of seeing who is coming in and out of a classroom or which teachers are "pulling" learners out of class. In this way, learners may feel more comfortable having conversations in a virtual setting.

We also want to acknowledge that the opposite may be true. For example, some learners may not have a place at home where they can speak openly or without disruption. Other learners may prefer to use the chat feature in the virtual platform versus having to unmute. This may be because they feel more comfortable communicating that way or because their family is also working virtually in the same room and it is noisy. When this happens, teachers may need to intentionally plan for their use during instruction.

In Figure 7.3, you will read a story from a teacher who worked through the barriers of communication while in the

"Most days that I teach, the majority of my learners do not have their cameras on. It was quite a learning curve to figure out other ways for me to get to know my students, as well as be able to gauge their thinking as we worked. The most common adaptation that I made was the use of the restrictions of the chat feature in Zoom. When we are working on tasks where I want the learners to provide their initial thinking, I often will set the chat feature so I am the only one to receive their response. This increases the level of participation because the learners did not feel they were so vulnerable in front of their classmates. Another one of my favorite adaptations that I have used, that I will continue to use is Flipgrid. The responses that I get through my learner's Flipgrid responses are authentic. I have a record of their thinking, as well as the vocabulary they are comfortable using. These are snapshots in time that can really capture their growth over time. I have not gotten to this place yet, but I would like to do a compilation of their videos over the year so they can see how their own thinking, reasoning, and vocabulary has grown over the course of the year."

FIGURE 7.3 A teacher's perspective on virtual learning

virtual environment. She mentions using Flipgrid, which is an online recording platform. It is easy to use and captures a student's response to a prompt within a designated time frame. Using this platform allows learners to record their thinking at a time that is convenient for them. As you read, consider how this teacher adapted to the virtual environment by varying her modes of communication with her learners.

The last big idea of Chapter 3 was getting to know your learners' mathematical knowledge through the use of wide tasks. Remember that wide tasks were defined as those that allow several possible strategies and may even have more than one correct solution. They are important because they support the beautiful interconnectedness of mathematics as described in the Five Strands of Mathematical Proficiency and position the learners as being capable and active in their learning.

The use of wide tasks is just as applicable in the virtual environment as in person. The trickier part of the virtual environment is digitally capturing the learner's thinking within wide tasks. Fortunately, many tools are available to support both teachers and students as they work within virtual Small Environments. We would like to highlight two tools that we have found to be particularly effective: Jamboard and Desmos. Both allow you to see a learner's work in real time as well as select and sequence learner work. We will elaborate on how to do this in the sections below.

Promoting Activity and the Need for Change in a Virtual Space

Chapter 4 focused on learners as active participants in the mathematics environment. We asked you to devote class time so learners may co-construct the social norms of the Small Environment. Although the Small Environment may now be a virtual space, it will be just as important, if not more so, for you to co-construct a visual *virtual* "looks like" and "sounds like" chart. We show an example in Figure 7.4. You might want to use a tool such as a Google Slide or Google Doc where learners are able to collaborate on the creation of this anchor chart, which can be referenced

In our classroom, when doing MATH

❑ Everyone participates	❑ Everyone shares their ideas
❑ Everyone listens to make sense of the speaker's words	❑ Everyone allows one person to talk at a time
❑ Everyone uses math tools appropriately	❑ Everyone uses kind and respectful words
❑ Everyone uses a raised hand to respond (real hand or icon)	❑ Everyone mutes themselves when they are not sharing
looks like	**sounds like**

FIGURE 7.4 Norms in the virtual space

and revised on a regular basis. An added benefit of creating these in a digital platform is that learners can find images to add to the norms chart to illustrate what it means to them to engage in mathematics. After the social norms have been co-constructed, just like in a non-virtual space, it may be beneficial to develop sentence starters or frames based on those norms for students to reference during the "Think-Pair-Share" structure, as described in Figure 7.5.

Chapter 4 also focused on the four classifications of tasks: Accessing, Bridging, Challenging, and Varying. As you design instruction for the virtual environment, you will maintain the intentional sequencing of these tasks, just as you would if you were in person. To think through the nuances, you might want to

Active listening & responding	Sharing your ideas
"I agree with you because…."	"I have an idea to share…"
"I respectfully disagree with you because…"	"I solved the problem by…"
"I like how you… Have you thought about…"	"I wonder if…"
"I understand how you…"	"I noticed that…"
"Can you explain how you…"	"First, I … Then I …"
"How did you know…"	

FIGURE 7.5 Sentence starter/sentence frames developed from norms

revisit Chapter 4 through the lens of the virtual environment. We will also unpack where to use the tasks when implementing the Think-Pair-Share structure in the virtual space below.

Promoting Noticing, Reflection, and Connections/ Conversation in a Virtual Space

Implementing the Think-Pair-Share structure, outlined in Chapter 5, in the virtual environment will have some amazing benefits but also some potential barriers. Using an online platform that allows you to monitor a learner's work in real time, such as those mentioned above, will be critical as you monitor learner thinking and incorporate prompts in order to advance their thinking. Let's unpack each part:

Think

In a virtual environment, a potential benefit is that some of the initial "Think" time may be available to learners asynchronously (i.e., outside of live instructional time). One way to accomplish this is to provide a wide task with the expectation that learners will work on it *outside of* class time. You might assign the task "Two students share five tacos equally. How much taco does each person get?" through one of the virtual platforms mentioned previously. As seen in Figure 7.6, this task was presented to learners through the Desmos platform. Desmos has drawing tools that allow learners to annotate their thinking. By presenting the task asynchronously, learners will have more time to think about the task(s). The benefit is that there is not the time pressure that there may be during synchronous class time. Another benefit of having the "Think" time completed asynchronously is that it provides more time for you to think about the intentional pairing of learners during the "Pair" part of the lesson structure.

However, providing tasks in an asynchronous space may mean that you miss out on monitoring learner thinking as the learner works through the problem. You will need to weigh the benefits of learners having additional "Think" time outside of class with the lost opportunity to monitor their work in real time.

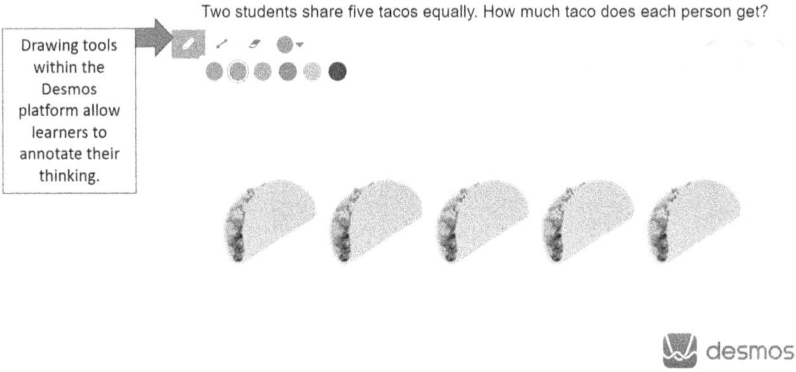

FIGURE 7.6 Student view of asynchronous task assigned in Desmos

Then determine, based on your learners' needs and the expected outcome of the lesson, which is the most efficient and beneficial use of your synchronous instructional time. Sometimes, you might offer them time to think outside of class time; other times, you want to monitor their initial thinking, therefore presenting the "Think" task during a synchronous session. Keep this in mind because both the "Pair" and "Share" parts are collaborative and will most likely happen in real time, so you will have the opportunity to monitor any adjustments in thinking during those parts.

Pair

In a virtual setting, we feel that the use of the breakout rooms is the most effective way to implement the "Pair" portion of the lesson because it allows in-the-moment dialogue. However, there must be intentionality with how breakout rooms are introduced and the expectations around them. As you see in Figure 7.7 there are three factors to consider: size of group, time in breakout room, and tasks used. During the first few experiences in breakout rooms, you will want to keep the group size to two people. This allows the learners to feel most comfortable and increases the likelihood that they will engage with the other person. If possible, you might want to consider pairing the students with someone you know they will talk to (a friend). You will also want to keep the amount of time very short (2 to 3 minutes) so the

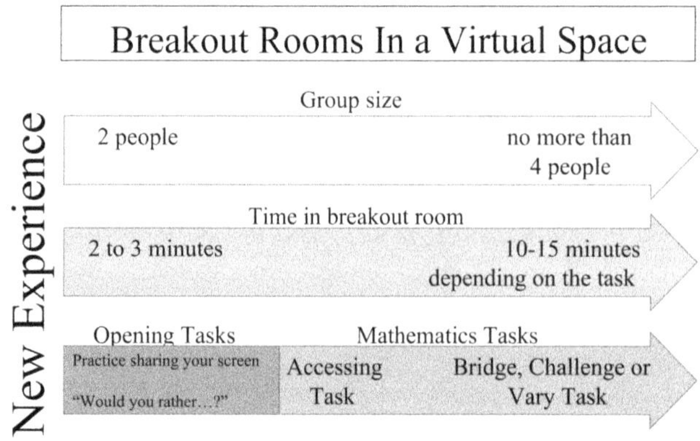

FIGURE 7.7 Guidance for breakout rooms

learners know they have a limited amount of time and need to stay on-task. The initial tasks may be of more of a technical nature, such as building their comfort with the platform (i.e., practice sharing the screen or using an annotation tool), or it might be engaging with a "low risk" question such as "Would you rather use a kangaroo or hippopotamus as a mode of transportation?" As learners' comfort level increases, you can increase the amount of time that they spend in the breakout room, you can increase the number of people in the group to three or four (although you may not do this in the Think-Pair-Share structure), and you can transition into mathematical tasks. Similar to what we outlined in Chapter 4, you will want to start with Accessing tasks, where there are multiple entry points that the learners can engage with. At this point, you will infuse Bridging, Challenging, and Varying tasks depending upon your overall goal for that lesson or sequence of lessons.

Although we feel that the use of the breakout rooms is the most effective way to implement the "Pair" portion of the lesson in the virtual setting, we understand that it may not be an option for everyone. If this is not feasible, the learners will need to communicate in written form or through the use of auditory or video recording (such as Flipgrid) to be shared with their partner as a means of dialogue. However, because it is not in real time, this might hinder the communication and comprehension between the learners.

Share

During the traditional (non-virtual) "Share" time in a lesson, a teacher needs to find a way to have learners share their work so that all students in class are able to view it. The problem is that it is typically on a piece of paper or a small dry-erase board. Some might project the students' work by using a document camera or they might have the students re-work their problem on the board. However, in a virtual environment share, all students are on their computer. As such, using platforms like Desmos and Jamboard (and others) allows all of the learners and teacher the opportunity to view the learners' work clearly because it is projected right to their screen. Another benefit of the use of these types of platforms is that the teacher can more easily sequence the learners' work for the "Share" part of the lesson. In Figure 7.8, you see an example of students' work when they are presented with this task through the Desmos platform: "Two students share five tacos equally. How much taco does each person get?" The teacher was able to monitor the students as they worked through the "Think" portion because their screen updates in real time. Then, based on what the teacher saw in their work as well as short conversations they had with each student in order to better understand their thinking, the teacher paired up Students A and C and Students B and D to discuss their thinking. While they were in their breakout rooms discussing their "Pair" portion of

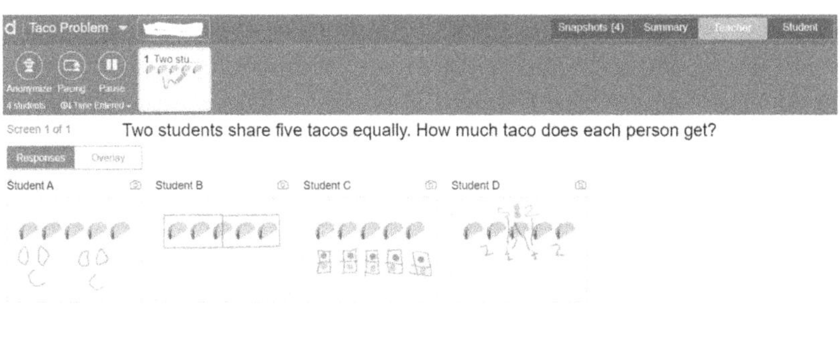

desmos

FIGURE 7.8 Teacher view in Desmos platform

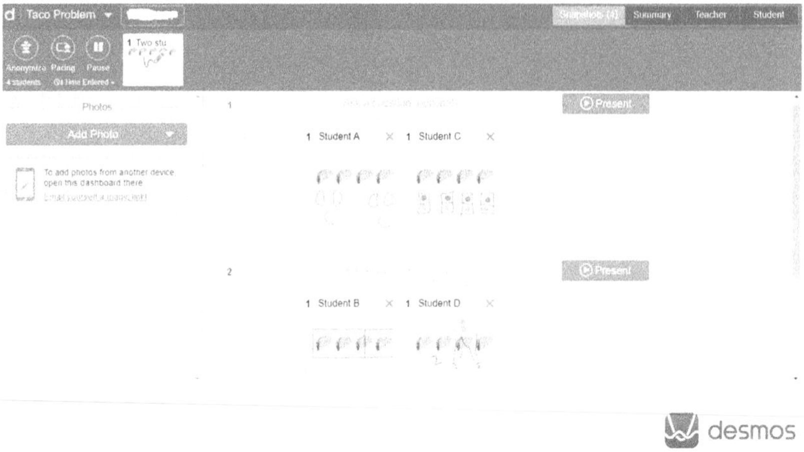

FIGURE 7.9 Snapshot feature in Desmos platform

the lesson, the teacher visited each of the breakout rooms. The teacher then used the "Snapshot" feature (see Figure 7.9 within Desmos to sequence the student work that was going to be shared with the whole group.

Revisiting Norms and Sociomathematical Norms in a Virtual Space

As we reflect about the Think-Pair-Share framework in the virtual space, let's revisit Table 5.5 – "Norms within Think, Pair, and Share" – from Chapter 5. We bring it to your attention again here in Table 7.2. How do social norms and sociomathematical norms need to be adjusted for the virtual space? Will you have the same social norms if your students are working synchronously as you would if they were working asynchronously? Revisit the questions listed with a virtual space in mind.

In summary, although the educational landscape has been forever changed, it has opened a new door that allows engagement and collection of learner thinking in ways that previously may have been left untapped. As we move forward, let's optimize learning by harnessing the best of both worlds and design effective learner-driven math interventions!

TABLE 7.2 Norms within Think, Pair, and Share

	Social norms	*Sociomathematical norms*
Think	• Do your students need absolute quiet "Think" time? • Do they need to talk out loud to themselves in order to process the information? • Is there a requirement to show your thinking or is it just personal reflection? • How will you navigate the students that want to just jump in and start talking? • What is the role of the teacher during "Think" time?	Students might ask themselves these questions: • How am I preparing my thinking to be shared with either a partner or the group? • What do I do if I get stuck? • Does my answer make sense?
Pair	• Who will determine who talks first? • What is the expected outcome of the conversation? • How will you ensure that students are actively listening and learning from each other?	Students might ask themselves these questions: • How does my thinking compare to my partner? • What are the similarities in our thinking? • What are the differences in our thinking? • Does their answer make sense?
Share	• How do you sequence the student thinking? • What is the expectation of the students during this time? Will they be expected to write anything down?	Students might ask themselves these questions: • How does my thinking compare with that of my peers? • What are the similarities in our thinking? • What are the differences in our thinking?

References

Carpenter, T., Fennema, E., Franke, M., Levi, L., & Empson, S. (2017). *Cognitively guided instruction*. New York, NY: Heinemann.

Desmos Platform Math Tools (2021). *Desmos Snapshot feature*. Retrieved January 12, 2021, from https://www.desmos.com/.

For Product Safety Concerns and Information please contact our EU representative GPSR@taylorandfrancis.com
Taylor & Francis Verlag GmbH, Kaufingerstraße 24, 80331 München, Germany

www.ingramcontent.com/pod-product-compliance
Ingram Content Group UK Ltd.
Pitfield, Milton Keynes, MK11 3LW, UK
UKHW022200190926
475702UK00003B/134